A Philosophical Case for Ecological Pessimism

Our current ecological crisis—featuring problems such as climate change, ocean acidification, and mass extinction—raises various moral issues, including a high probability of injustice and massive harm. This book defends a position called ecological pessimism, an attitude whose core feature is the belief that ecological catastrophe is likely to occur in the future.

The author's defense of ecological pessimism has two components. First, he makes the case that the relevant ecological facts about our world make ecological pessimism a reasonable, and indeed plausible, expectation. Second, he argues that ecological pessimism is morally and practically appropriate. Ecological pessimism is a distinctively moral kind of pessimism because the failure to avert ecological catastrophe leads to great ills for human beings and non-human nature. The author's account responds to likely objections to ecological pessimism and makes the case against ecological optimism. Despite this, the author makes clear that being pessimistic about our ecological prospects is compatible with the melioristic project of improving our bad condition. He argues that environmental philosophy as a way of life, with its emphasis on environmental virtue and rich resources for developing spiritual exercises, is both a robust and attractive option for an ecological pessimist.

A Philosophical Case for Ecological Pessimism will appeal to scholars and graduate students working on ethics and environmental philosophy.

Toby Svoboda teaches philosophy and environmental studies at Colgate University. His previous book is *A Philosophical Defense of Misanthropy* (Routledge, 2022).

A Philosophical Case for Ecological Pessimism

Toby Svoboda

Routledge
Taylor & Francis Group

NEW YORK AND LONDON

First published 2025
by Routledge
605 Third Avenue, New York, NY 10158

and by Routledge
4 Park Square, Milton Park, Abingdon, Oxon, OX14 4RN

Routledge is an imprint of the Taylor & Francis Group, an informa business

© 2025 Toby Svoboda

The right of Toby Svoboda to be identified as author of this work
has been asserted in accordance with sections 77 and 78 of the
Copyright, Designs and Patents Act 1988.

All rights reserved. No part of this book may be reprinted or
reproduced or utilised in any form or by any electronic, mechanical,
or other means, now known or hereafter invented, including
photocopying and recording, or in any information storage or
retrieval system, without permission in writing from the publishers.

Trademark notice: Product or corporate names may be trademarks
or registered trademarks, and are used only for identification and
explanation without intent to infringe.

ISBN: 978-1-032-94401-2 (hbk)
ISBN: 978-1-032-94399-2 (pbk)
ISBN: 978-1-003-57052-3 (ebk)

DOI: 10.4324/9781003570523

Typeset in Sabon
by Deanta Global Publishing Services, Chennai, India

To Hogan and Paul

Contents

Acknowledgments

The author thanks Allen Thompson, Kate Woodford, Annabell Woodford, and several anonymous reviewers, as well as audiences at Colgate University, John Caroll University, the International Association for the Study of Environment, Space and Place, and the International Society for Environmental Ethics for feedback on draft chapters of this project. The author acknowledges permission from Brill Publishers and Indiana University Press to adapt material previously published in *Contemporary Pragmatism* and *Ethics & the Environment*.

Introduction

This book defends what I call ecological pessimism, an attitude whose core feature is the belief that ecological catastrophe is likely to occur in the future, a kind of view that has rarely been defended.[1] Broadly speaking, my defense has two components. First, I make the case that the relevant ecological facts about our world make this a reasonable, and indeed plausible, expectation. Second, I argue that ecological pessimism is morally and practically appropriate.

Pessimism and Misanthropy

In some ways, this book is a follow-up to my previous work, *A Philosophical Defense of Misanthropy*, although the present book in no way requires familiarity with the earlier one. A brief summary will suffice. In the earlier book, I argued that humanity has been a moral disaster. History and current affairs provide ample evidence for this: slavery, genocide, wars of aggression, cruelty to animals, and environmental destruction. This warrants the belief that humanity in general is morally bad. Because this type of misanthropy is belief-based rather than desire-based, it is possible to assess it in epistemic terms. As I try to show, this misanthropy is plausible, justified, and likely true. The case for misanthropy depends upon a kind of inductive argument, drawing evidence from historical, present-day, and likely future atrocities. Of course, some human beings are morally decent, even if that is a rare thing. In order to account for this, I defend an asymmetry thesis, according to which moral ills are significantly more important than moral goods. Consider the murderous philanthropist (or philanthropic murderer, if you like), who saves a thousand lives while taking "only" ten. This is just a rough sketch, and anyone who is interested may consult the full arguments in the book itself.

As with my defense of misanthropy, my defense of ecological pessimism takes a dim view of humanity. Both are tied up with a kind of moral pessimism about humanity's future, holding that we are unlikely to make

DOI: 10.4324/9781003570523-1

serious moral progress as a species. Further, like the judgment that humanity is morally bad, the judgment that ecological catastrophe is likely is belief-based rather than desire-based, allowing for an epistemic evaluation in terms of its truth and justification, as well as the embedding of that judgment in chains of reasoning. However, accepting ecological pessimism does not require accepting my version of misanthropy. One might think (falsely) that human beings are morally wonderful creatures while consistently holding that ecological catastrophe is indeed likely to occur. For example, despite our best moral efforts, avoiding catastrophe could prove to be beyond our abilities. That would be tragic rather than morally obscene. This shows that the positions of misanthropy and ecological pessimism can come apart.

Admittedly, my own account of why ecological catastrophe is likely does appeal to the moral ills of humanity in two ways. First, moral ills like injustice and harm are partly what make certain outcomes catastrophic. On a planet devoid of life, ocean acidification would be merely an interesting phenomenon to observe from afar. On our planet, it is causing extensive harm to human communities, non-human organisms, and marine ecosystems. Dangerous climate change is bad in part because of the evident injustice of its likely impacts, burdening low-emitters with disproportionate and undeserved harm. Second, on my account, humanity's moral failings partly explain why ecological catastrophe is likely. We could do much more to reduce the many risks associated with climate change, ocean acidification, mass extinction, and the like. We do little despite the fact that, to a large extent, some segments of humanity are morally responsible for these ecological problems. These misgivings about humanity fit well with the misanthropic stance of the earlier book, but the arguments in this book stand or fall on their own. There is no need to consult the misanthropy book in order to fully assess the current one.

Chapter Overview

Chapter 1: Ecological Risks

The opening chapter examines our current ecological crisis, describing phenomena such as climate change, ocean acidification, mass extinction, and the potential use of nuclear weapons. Many of these are familiar, but it is necessary to describe them in some detail in order to prepare the ground for my main arguments in subsequent chapters. I devote the most attention to climate change, discussing its potential impacts, "tipping points" in the climate system, and the prospect of geoengineering. The chapter also distinguishes between mere events and genuine catastrophes. The ecological risks discussed are potentially catastrophic because of their impacts on human and non-human subjects. For those who are already familiar with these risks, this chapter may be skipped without much loss.

Chapter 2: Understanding Pessimism

Chapter 2 aims to clarify what I take the concept of ecological pessimism to involve. It is the expectation that the future will involve ecological catastrophes of some kind. I offer an account of what qualifies as a catastrophe, and I specify what sort of attitude constitutes an expectation in the relevant sense. I take this attitude to be cognitive in nature, which is to say that it is a belief rather than some desire-like or emotive attitude. Of course, this cognitivist expectation might be accompanied by various non-cognitive attitudes, such as fear, anxiety, apathy, and so on, but these do not constitute the core of what I understand to be the ecologically pessimistic attitude. Further, ecological pessimism is a distinctively moral kind of pessimism, for it is rooted in the belief that humanity will fail to avert avoidable ecological catastrophe. The specific targets of ecological pessimism include harm to sentient entities and injustice to human beings.

Chapter 3: Evil in Environmental Affairs

Some of the actions that make ecological catastrophe likely are not only morally questionable—they are genuinely evil. This might sound implausibly extreme, but there is a very strong case for it. An ecological catastrophe, say one driven by a runaway greenhouse effect, would devastate life on earth, both human and non-human. We reserve judgments of evil for the most heinous of crimes, such as mass genocide. The suffering and injustice of an ecological catastrophe would likely be much greater than that of other moral ills that we plausibly identify as evil. A good example of this is what I call climate obstructionism or any coordinated attempt to hinder, slow, or undermine progress in addressing the crisis of climate change. Such obstructionism can take many forms: denying the reality that the climate is changing, manufacturing doubt regarding the anthropogenic nature of climate change, downplaying the severity or probability of potential impacts, lying about the costs of renewable energy, and many more. Those who engage in this obstruction, as opposed to some of their targets, are educated individuals who are well aware of the reality and potential damage of climate change. Climate obstructionism is morally reprehensible to the point of being genuinely evil. My argument for this will come in two parts. First, I sketch what plausibly counts as evil in the relevant sense. Second, I show that climate obstructionism satisfies reasonable criteria for qualifying as evil in the specified sense. I also respond to the concern that employing judgments of evil is dangerous due to its potential misuse in social and political contexts. Finally, this chapter addresses the Nietzschean critique that charges of evil are motivated by *ressentiment*.

Chapter 4: Objections to Ecological Pessimism

This chapter considers likely objections to the ecological pessimism I defend. Such objections include the following: there is deep uncertainty about the future, and therefore we lack sufficient justification for any expectation of ecological catastrophe; ecological pessimism is unreasonably alarmist; there are various pragmatic problems with such pessimism, including the possibility that it could undermine the motivation to work toward environmental progress; and the charge that my view misses the mark, because the supposed "ecological" catastrophe I expect is really only a catastrophe for certain species, including humanity. I respond to each of these objections, showing that they do not succeed in undermining the case for ecological pessimism.

Chapter 5: Environmental Philosophy as a Way of Life

Despite my pessimism, environmental philosophy can still play a valuable role. Here I defend the idea of environmental philosophy as a way of life, modeled on the approach of ancient schools of philosophy, such as Epicureanism and Stoicism. This approach identifies an environmental conception of the good life and spiritual exercises meant to help an individual realize that good life. One reason this type of environmental philosophy is worth pursuing is that it offers a way for the ecological pessimist to live well despite her pessimism. Ecological catastrophe may be likely, but this need not prevent us from living good lives, particularly when the good life is understood to involve environmental virtues that we can cultivate.

Chapter 6: A Case for Meliorism

This chapter builds on my response to one of the objections considered in Chapter 4. Being pessimistic about our ecological prospects is compatible with the melioristic project of improving our bad condition. Pessimism does not entail quietism. I draw on American pragmatism, such as the work of Charles Sanders Peirce and William James, for my understanding of what meliorism involves. There are several reasons why we should aim to reduce ecological risk, even if doing so will not avert catastrophe. First, there is practical value in mitigating ecological ills, because that can at least reduce the damage that those ills would otherwise bring. Second, we have a moral obligation to meliorate ecological ills, because doing so can reduce the unjust harm that relevant parties would otherwise experience.

Chapter 7: Why Not Optimism?

This last chapter finally makes the case against ecological optimism, which we may define as the expectation that we will likely avoid ecological catastrophe in the future. I begin with two main arguments. First, such optimism is not plausible, given the many environmental problems described in Chapter 1 and our dismal track record in responding to them. Second, such optimism is dishonest, because it obscures the ugly truth regarding humanity's impact on ecological systems. This leads to a discussion of the pragmatic problems with ecological optimism. If taken seriously, it has the potential to create false hopes and unrealistic expectations. This may take attention away from the melioration that would be possible under a more realistic, and hence pessimistic, attitude.

Note

1 For an interesting exception, see Nguyen (2024).

Reference

Nguyen, Anh-Quân. "Pessimism for Climate Activists." *Ethics & the Environment* 29, no. 1 (2024): 109–137.

1   Ecological Risks

Our environmental prospects are bleak. Despite decades of warnings from relevant experts, humanity has done relatively little to reduce risks of ecological catastrophe. There are many such risks associated with climate change, ocean acidification, species extinction, and nuclear war, for example. In this chapter, I will offer a brief overview of these risks. Although certain potential impacts are intuitively catastrophic, I will save direct discussion of what counts as a genuine catastrophe for the next chapter.

Climate Change

Dangerous climate change includes many ecological risks. We have long understood the greenhouse effect and have been aware of increased concentrations of greenhouse gasses in the atmosphere (Arrhenius 1896; Keeling 1961).

Impacts

The dangerous effects of climate change have been evident since at least the first assessment report of the Intergovernmental Panel on Climate Change (IPCC) in 1990. That report issued the following warning:

> Based on the existing literature, the studies have used several scenarios to assess the potential impacts of climate change. These have the features of: (i) an effective doubling of CO2 in the atmosphere between now and 2025 to 2050 for a 'business-as-usual' scenario; (ii) a consequent increase of global mean temperature in the range of 1.5°C to 4°-5°C; (iii) an unequal global distribution of this temperature increase, namely a smaller increase of half the global mean in the tropical regions and a larger increase of twice the global mean in the polar regions; and (iv) a sea-level rise of about 0.3–0.5 m by 2050 and about 1 m by 2100,

DOI: 10.4324/9781003570523-2

together with a rise in the temperature of the surface ocean layer of between 0.2° and 2.5°C.

(IPCC 1990, 1)

These are not mild changes. One might expect a warning like this to spur decisive action among policymakers and their constituents, but that has not been the case. Clearly, there has been some movement in the right direction—the formation of the United Nations Framework Convention on Climate Change (UNFCCC), emissions mitigation efforts in some countries, and the growth of renewable energy—but it falls far short of what is needed.

Climate change promises to be very harmful to human beings, non-human organisms, and natural systems. Much of this harm will fall short of catastrophe. In the cautious words of a more recent IPCC report:

> Human-induced climate change, including more frequent and intense extreme events, has caused widespread adverse impacts and related losses and damages to nature and people, beyond natural climate variability. [] Across sectors and regions the most vulnerable people and systems are observed to be disproportionately affected. The rise in weather and climate extremes has led to some irreversible impacts as natural and human systems are pushed beyond their ability to adapt.

(IPCC 2022, 9)

The IPCC goes on to note that climate change is causing "increasingly irreversible losses, in terrestrial, freshwater and coastal and open ocean marine ecosystems," and to a greater degree than thought in previous reports. The impacts on ecosystems have brought "adverse socioeconomic consequences. About half of assessed species have migrated to higher latitudes or altitudes. Species have been negatively affected by extreme heat and "mass mortality events." Some of these events are irreversible, such as initial extinctions, while others "are approaching irreversibility such as the impacts of hydrological changes resulting from the retreat of glaciers, or the changes in some mountain … and Arctic ecosystems driven by permafrost thaw" (IPCC 2022, 9).

As the report continues, climate change is "contributing to humanitarian crises." For example,

> Climate and weather extremes are increasingly driving displacement in all regions (high confidence), with Small Island States disproportionately affected (high confidence). Flood and drought-related acute food insecurity and malnutrition have increased in Africa (high confidence) and Central and South America (high confidence).

Similarly, "Through displacement and involuntary migration from extreme weather and climate events, climate change has generated and perpetuated vulnerability (medium confidence)" (IPCC 2022, 11). The IPCC expects risks to increase in the future: "For 127 identified key risks, assessed mid- and long-term impacts are up to multiple times higher than currently observed (high confidence)." This depends on both the extent of warming and the degree to which human communities adapt, but "projected adverse impacts and related losses and damages escalate with every increment of global warming (very high confidence)" (IPCC 2022, 12). At the same time, these various risks and impacts "are becoming increasingly complex and more difficult to manage. Multiple climate hazards will occur simultaneously, and multiple climatic and non-climatic risks will interact, resulting in compounding overall risk and risks cascading across sectors and regions" (IPCC 2022, 18). Unfortunately, there are limits to human adaptation. Some so-called "soft limits" have already been reached, although these limits "can be overcome by addressing a range of constraints, primarily financial, governance, institutional and policy constraints (high confidence)." Unfortunately, "hard limits to adaptation have been reached in some ecosystems (high confidence). With increasing global warming, losses and damages will increase and additional human and natural systems will reach adaptation limits (high confidence)" (IPCC 2022, 26).

Although these projections are worrisome enough, climate prospects may be worse than IPCC reports suggest, perhaps far worse. Quite reasonably, the IPCC operates on a sort of consensus model, drawing upon a large body of research to produce a report every few years. This tends to downplay the possibility of extreme outcomes, which may be dismissed as outliers. Yet it could turn out that a current scientific consensus is mistaken, perhaps because that consensus was reached prematurely (Oppenheimer et al. 2007).[1] One possible example is consensus around climate sensitivity, which is defined as the global temperature increase that would occur if atmospheric greenhouse gas concentrations doubled relative to pre-industrial concentrations. Much depends on the correct value for climate sensitivity, but there is uncertainty here (Sherwood et al. 2020; Roe and Baker 2007). A value of 2°C or less would make continued emissions substantially less risky than would a value of 5°C or greater. As John Broome notes, the IPCC has acknowledged a greater than 5 percent probability that climate sensitivity is up to six degrees, with a smaller probability that it could be as high as ten degrees: "Ten degrees of warming would be a great catastrophe. It would cause dreadful destruction and suffering. It would also entail a collapse of the planet's human population" (Broome 2012, 130–131). This possibility raises challenges for modeling the economics of climate change (Weitzman 2009). The takeaway from all this is to say that climate catastrophe is a realistic possibility, even if it is relatively unlikely. This suggests that we should pay more attention to that possibility (Davidson and Kemp 2024; Kemp et al. 2022).

Tipping Points

A tipping point is understood to be a point at which a small change to a system leads to a disproportionate and possibly irreversible change in that system. Imagine a ball poised at the top of a cliff. A brief gust of wind might "tip" the ball over the edge, causing it to fall a great distance. Returning the ball to the top of the cliff would require much more energy than was expended in tipping it over the edge and is very unlikely to occur naturally. I will mention two types of potential tipping points in the climate system.

First, the Atlantic Meridional Overturning Circulation (AMOC) is a "conveyor belt" of ocean currents, transferring warm water to higher latitudes and cold water to lower latitudes in the Atlantic Ocean. It is crucial for maintaining regional climates, including that of northwestern Europe. One expected effect of climate change is a weakening, and potentially a full shutdown, of the AMOC. This is supported by various modeling studies. There is evidence that the AMOC has already weakened substantially during the past century (Rahmstorf et al. 2015). This could have unwelcome impacts. According to one study, a weakened AMOC could result in

> widespread cooling throughout the North Atlantic and northern hemisphere in general; less precipitation in the northern hemisphere midlatitudes; large changes in precipitation in the tropics and a strengthening of the North Atlantic storm track. The general cooling and atmospheric circulation changes result in weaker peak river flows and vegetation productivity, which may raise issues of water availability and crop production.
>
> (Jackson et al. 2015)

There is uncertainty surrounding the timing of a full shutdown, but it could occur as soon as mid-century (Ditlevsen and Ditlevsen 2023).

Second, ice sheet melting is a major contributor to sea-level rise. The West Antarctic and Greenland ice sheets contain massive quantities of freshwater. Due to oceanic and atmospheric warming, each is at risk of collapsing, which would greatly exacerbate sea-level rise (IMBIE, 2020; Pan et al. 2021). In the case of the West Antarctic Ice Sheet, it may be already too late to avert its collapse via realistic emissions mitigation scenarios, which is to say that the tipping point may already have been reached (Naughten, Hollan, and Rydt 2023). As with the AMOC, there is extensive uncertainty here, given the dynamic nature of ice sheets (Noble et al. 2020; Sadai et al. 2020). Nonetheless, it is clear that a collapse of one or both of these major ice sheets is a realistic possibility, as well as that such a collapse would greatly increase sea level. There is evidence that even 1.5°C of warming may trigger various tipping points in the climate (Armstrong McKay et al. 2022), many of which could prove to be catastrophic.

Geoengineering

One indication of our dire, or soon-to-be-dire, climate situation is that some respectable scientists are giving serious consideration to geoengineering as a response to climate change. Defined as the intentional, large-scale, technological modification of the global environment (Keith 2000), geoengineering is usually divided into two categories. First, carbon dioxide removal (CDR) includes techniques that aim to reduce atmospheric concentrations of greenhouse gasses, including direct air capture and ocean iron fertilization (Keller et al. 2018). If they work, such technologies would pull greenhouse gasses out of the atmosphere and sequester them elsewhere, such as underground or in the deep ocean. The point of this would be to reduce global warming, which is driven by increased concentrations of atmospheric greenhouse gasses.

The second category, solar radiation management (SRM), is more controversial. Rather than drawing down atmospheric greenhouse gasses, as CDR would do, SRM techniques seek to cool the planet artificially, thereby compensating for the warming of greenhouse gasses. Possible techniques include marine cloud brightening and stratospheric aerosol injections (SAI), the latter of which has received the most attention. SAI would mimic the effects of a large volcanic eruption by injecting into the upper atmosphere several megatons of sulfates, which have the property of reflecting incoming solar radiation management. Roughly put, the aim of this would be to increase the reflectivity of the stratosphere, reflecting some quantity of sunlight. By reducing the amount of energy available for absorption by the planet, this could induce a degree of cooling that compensates for some or all of the warming caused by anthropogenic greenhouse gas emissions. This is not the place to offer a more thorough, technical account of SAI, but the foregoing will suffice.

Over the past fifteen years or so, SAI has come under serious consideration by respected scientists, where previously it was a forbidden topic (Boettcher and Schäfer 2017; Crutzen 2006). The reason for this is readily acknowledged by proponents of researching SAI, namely the failure of the global community to address climate change in a serious way. Virtually all advocates of SAI research—virtually no one advocates immediate or near-term deployment of SAI—agree that emissions cuts would be preferable to geoengineering. The problem, of course, is that humanity has failed to do this to a sufficient degree, and the future is not promising on that front. Thus some have begun looking into other options, including some that were once unthinkable. Let me stress that we should resist the temptation to turn SAI researchers into strawmen. Relying on both their published views and my conversations with many of them, these individuals are not arrogant Prometheans who gleefully aim to reshape the planet. Admittedly, some past scientists could be accused of holding that attitude (see Fleming 2010), but the fantasy of improving nature through technological intervention has been discredited. Obviously, deploying or even researching SAI might still be a terrible idea, but we should critique SAI proponents for their actual views and proposals, not imaginary ones.

Unsurprisingly, the growing interest in SAI has alarmed many, including scientists, activists, and ethicists. The potential problems of injecting aerosols into the stratosphere are many: ozone depletion, changes in regional precipitation patterns, unaddressed ocean acidification, unilateral deployment by one state against the wishes of others, the prospect of premature termination, and more (Robock 2008). Each of these could cause serious environmental damage, harm to humans and non-humans alike, and injustice to some persons or groups (Svoboda et al. 2011). Many commentators have critiqued SAI on ethical grounds (Pamplany et al. 2020). Again, I cannot go into detail on all of these, but let us consider the case of the so-called termination problem.

Sulfate aerosols have a stratospheric lifetime of about three to five years, so in order to maintain a constant cooling effect—say, a reduction of 2°C—stratospheric aerosols would need to be carefully monitored and constantly resupplied, which might be done through the use of high-altitude aircrafts, balloons, or other means (Smith et al. 2018). Should this cease, models show that the global average temperature would rise rapidly, given the warming effect of atmospheric greenhouse gasses (Matthews et al. 2007). Two degrees of warming over the course of five years would be much more damaging than the fairly gradual warming we are currently experiencing, which itself is already causing substantial damage. Although the exact effects of premature SAI termination are unclear, we can plausibly view it as a catastrophe. Of course, one might say this incentivizes us not to discontinue SAI once it is started, but there will always be some risk of termination. Baum et al. (2013) consider the idea of "double catastrophe," in which one catastrophic event (e.g., a pandemic or nuclear war) interferes with our ability to maintain SAI, thus leading to the second catastrophe of premature termination.

The implications are troubling. One might try to justify SAI as a means of avoiding catastrophic climate change. Indeed, this is usually the framing that is most friendly to SAI, viewing it as the lesser of two evils or as an emergency intervention (Svoboda 2017). Yet deploying SAI creates a new risk of catastrophe, namely premature termination. This would be an odd state of affairs. In order to avoid catastrophe, we might put ourselves at risk of a new catastrophe. This looks absurd, but we have demonstrated little interest in safer and more ethical policies, such as cutting emissions. It is tempting to blame this on powerful special interests, but millions of ordinary citizens support inaction on climate change by voting for climate obstructionists. It is true that most citizens do not support such politicians *because* of the latter's climate obstructionism. Rather, many people vote on the basis of irrational fears about (say) immigration. Yet the effect is the same. Very few people actually care about the threat to organized life on earth posed by climate change. There is little reason to expect this to change, so it is plausible to think that we will face a dilemma: either risk the catastrophe of unchecked climate change or risk the catastrophe of prematurely terminated SAI. This all provides ample and reasonable motivation for pessimism, as we shall see throughout this book.

Other Ecological Risks

I will now discuss several other ecological risks, albeit more briefly than in the case of climate change, namely ocean acidification, species extinction, and nuclear war. I focus on these because they are understood fairly well and pose realistic threats to human and non-human beings, whereas many other ecological risks (e.g., from artificial intelligence) are highly speculative at this stage.

Ocean Acidification[2]

Ocean acidification is distinct from climate change, although the two are causally related at the present time. The former concerns changes in ocean chemistry, whereas the latter concerns changes in average weather patterns and their distributions. They are causally related because anthropogenic emissions of CO_2 are driving both processes. Ocean acidification is caused by increased concentrations of CO_2 in the atmosphere, which is absorbed by the oceans and reacts with other chemicals to produce positively charged hydrogen ions ($H+$), which in turn reduce ocean pH and upset the chemical balance of ocean ecosystems (Doney et al. 2009). Ocean pH has decreased since the beginning of the Industrial Revolution and is expected to decrease further as atmospheric CO_2 continues to accumulate (Orr et al. 2005). This change in ocean chemistry affects marine calcifying organisms, such as corals, that rely on carbonate ions ($CO_{3}2-$) to form shells of calcium carbonate ($CaCO_3$) (Cornwall et al. 2021), although some types of calcifier may be less affected than others (Leung et al. 2022). CO_2 reacts with carbonate ions ($CO_{3}2-$) and water to form bicarbonate ions ($HCO_{3}-$), thus reducing the amount of carbonate available for calcifying organisms (Orr et al. 2005). In short, many marine organisms need carbonate ions, but an increased abundance of CO2 reduces the availability of carbonate ions.

It is not known with certainty to what degree further ocean acidification would affect marine calcifying organisms, in part because it is unknown how well these organisms can adapt to such changes (Doney et al. 2009). However, Orr et al. (2005) find that the shells of live pteropods exposed to decreased levels of carbonate ions exhibit "notable dissolution." Moreover, increased ocean acidification threatens the viability of entire coral reef ecosystems (Hoegh- Guldberg et al. 2007), because high levels of CO2 can kill corals and other organisms that are essential to such ecosystems. Although more research is needed on the degree and extent of harm likely to be caused by ocean acidification, it clearly poses a threat to marine ecosystems. If current emissions trends continue, the world's oceans will become more acidic.

One might ask what is so bad about oceans with a lower pH. First, ocean acidification has the potential to cause serious harm to non-humans. It threatens the very existence of calcifying marine organisms and thus the existence and well-being of those organisms that rely on calcifying organisms for food. Ocean acidification also threatens the very existence of coral reefs and thus the existence and well-being of those organisms that rely on coral reef ecosystems for habitat. The destruction of coral reef ecosystems or the depletion of various marine organisms poses threats to fisheries and tourism, the latter of which is often a vital source of income for many humans in developing countries. Further, coral reefs provide valuable ecosystem services for humans, such as coastal protection (Doney et al. 2020; Hoegh-Guldberg et al. 2007).

Species Extinction

Extensive species extinction is likewise distinct from climate change, although the latter is a major driver of the former at the present time (Román-Palacios et al. 2020). While there is some debate as to whether the current loss of species technically qualifies as the planet's sixth mass extinction (Kaiho 2022; Cowie et al. 2022), it is undeniable that species are disappearing at an alarming rate, far above the background rates of both prehistory and early human history (Rounsevell et al. 2020). In many cases, the disappearing species are unknown to science. There is evidence that undescribed species are going extinct at a higher rate than described ones (Liu et al. 2022). By most indications, we should expect extinctions to accelerate. Once again, while species extinction is distinct from climate change and ocean acidification, both are contributors to it. Other major causes include habitat loss, pollution, and the spread of invasive species (Gonçalves-Souza et al. 2020; Hogue and Greon 2022). Virtually all of these causes are ultimately anthropogenic. It is due to human activity that the climate is changing, that the oceans are growing more acidic, that habitat is destroyed, that air and waterways are polluted, that invasive species are spread around the world, and so on.

Of course, all this has ramifications for the species that manage to avoid extinction, including humanity. As one study finds:

> The ongoing sixth mass extinction may be the most serious environmental threat to the persistence of civilization, because it is irreversible. Thousands of populations of critically endangered vertebrate animal species have been lost in a century, indicating that the sixth mass extinction is human caused and accelerating. The acceleration of the extinction crisis is certain because of the still fast growth in human numbers and consumption rates. In addition, species are links in ecosystems, and, as they fall out, the species they interact with are likely to go also. In the regions where disappearing species are concentrated, regional biodiversity collapses are likely occurring. Our results reemphasize the extreme urgency of taking massive global actions to save humanity's crucial life-support systems.
>
> (Ceballos et al. 2020)

Reasonably enough, this study views current extinctions as constituting a threat to humanity, but they also constitute a threat to non-humans, such as death and suffering for sentient animals affected by the loss of food sources or habitat. Even if we limit consideration to humanity, the current extinction event poses a threat to human civilization, which relies on so-called "ecosystem services" ultimately provided by disappearing species: "Ecosystem functioning, including primary productivity, the biogeochemical cycles, and the network of trophic mutualistic and antagonistic species interactions that compose the food chains, is the fabric of life—a fabric that is translated by humans as ecosystem services" (Dirzo, Ceballos, and Ehrlich, 2022). Such services include the provision of food, pollination in agriculture, purification of air and water, aesthetic gratification, recreation, and many more. Accordingly, large-scale extinction could lead to reduced agricultural productivity, the collapse of food sources (e.g., fisheries), the loss of aesthetic marvels, and so on. As with the other risks discussed in this chapter, there is a great deal of uncertainty regarding the exact nature and timing of specific impacts, but it is very clear that species extinction poses a serious threat to human and non-human welfare.

Nuclear War

Nuclear war is not often discussed in an environmental context, but it poses a serious ecological risk. In addition to being devastating for human individuals and systems, a large-scale nuclear exchange would likely cause extensive damage to non-human organisms and systems. Obviously, a nuclear detonation would damage or destroy organisms and ecosystems in its immediate vicinity, but a nuclear war could also have global environmental impacts through the phenomenon of "nuclear winter" (Baum 2015; Robock et al. 2007). As Alan Robock writes:

> Nuclear winter is the term for a theory describing the climatic effects of nuclear war. Smoke from the fires started by nuclear weapons, especially the black, sooty smoke from cities and industrial facilities, would be heated by the Sun, lofted into the upper stratosphere, and spread globally, lasting for years. The resulting cool, dark, dry conditions at Earth's surface would prevent crop growth for at least one growing season, resulting in mass starvation over most of the world. [...] More people could die in the noncombatant countries than in those where the bombs were dropped, because of these indirect effects.
>
> (Robock 2010)

Like most studies on the topic, Robock focuses on the impacts for humanity, but it is clear that nuclear winter would be harmful to many other species. The same global cooling and darkening that devastates agriculture would likely devastate wildlife as well, for example, by inhibiting plant growth and thereby impacting animals that depend upon the inhibited plant species. A nuclear war could have additional ecological impacts, such as "wildfires, radioactive fallout, enhanced ultraviolet radiation, loss of atmospheric oxygen, gain in atmospheric carbon dioxide, and reductions in sunlight and temperature" (Westing 1987). The use of nuclear weapons could alter the planet's oceans. One modeling study found the following:

> Phytoplankton production and community structure are highly modified by perturbations to light, temperature, and nutrients, resulting in initial decimation of production, especially at high latitudes. A new physical and biogeochemical ocean state results... In the largest US-Russia scenario (150 Tg), ocean recovery is likely on the order of decades at the surface and hundreds of years at depth, while changes to Arctic sea-ice will likely last thousands of years, effectively a "Nuclear Little Ice Age." Marine ecosystems would be highly disrupted by both the initial perturbation and in the new ocean state, resulting in long-term, global impacts to ecosystem services such as fisheries.
>
> (Harrison et al. 2022)

Yet again, there is much uncertainty when it comes to the exact impacts of any particular scenario involving the use of nuclear weapons, but it is obvious that a large-scale nuclear war would be catastrophic in many ways, one of those being its impact on non-human nature.

It would be a mistake to dismiss the risk of nuclear war as archaic as if it were nothing more than a Cold War relic. In fact, tensions between nuclear powers have been rising in recent years (Brands and Gaddis 2021), and the proliferation of nuclear weapons remains a concern (Herzog 2020), even if proliferation has taken on a new form in recent times (Kaplow 2024). Appealing to the Cold War doctrine of "mutual assured destruction" (MAD, appropriately), it is sometimes claimed that a first use of nuclear weapons would be irrational, as it would ensure the first users' own destruction (Sokolski 2004). If that is true, then we should be fine so long as nuclear powers always behave rationally. This is not reassuring. If it was not already obvious, the last few years have shown that leaders of nuclear-armed states can be very stupid and reckless. Leaving the issue of rationality aside, MAD might not deter religious fanatics, and it provides little protection against an accidental launch. Indeed, there have been numerous close calls in the past, in which disaster was averted thanks to the decision-making of certain individuals (Tetrais 2017). While some take this to be evidence of "human prudence and the efficiency of mechanisms devoted to the guardianship of nuclear weapons" (Tetrais 2017), it only takes one such mistake for catastrophe to unfold: a panicked decision, a lapse of judgment, a drunk president, false information, incompetence somewhere in a chain of command, or whatever. In any given instance, there may be a low probability of nuclear war, but we continue to run that risk over many instances. We have avoided disaster over the past few decades. We can hope that our luck holds over the coming centuries, but there is obviously no guarantee.

Events or Catastrophes?

Taken by itself, there is nothing necessarily catastrophic about even extreme climate change, ocean acidification, species extinction, or nuclear detonations. These are merely geophysical processes. If we observe a distant, uninhabited planet being obliterated by a supernova, we are unlikely to describe the event as catastrophic. Because there is no life on the planet, there are no subjects that could be harmed by the supernova. There will be no suffering imposed nor any biotic process thwarted. One might say that the supernova is catastrophic for the planet itself, but this is likely to be used in a metaphorical sense. Again, the planet itself does not have a good. It is not the sort of thing that can be harmed or benefited by an astrological event. Now one could adopt the uncommon view that a planet does have a good in some sense and that the supernova harms the planet by interfering with that good. I will not claim to refute this possible view, but very few are likely to adopt it. Setting aside this unlikely exception, we are most likely to view the planet's destruction as an occurrence that is interesting to us in some way, say scientifically or aesthetically, but not as a catastrophe. Now if that same planet happened to be inhabited by human-like and animal beings who experienced a great deal of pain during the event, we would be compelled to change our view. This would clearly count as a catastrophe for the inhabitants of the destroyed planet.

To call something "catastrophic" is to issue a value judgment, taking some change to be bad to a substantial degree. In the next chapter, I will offer more clarification about what counts as an ecological catastrophe, but for now, I wish only to make the point that geophysical changes are taken to be bad or good only in relation to things for which such changes matter. In particular, a catastrophe needs to be catastrophic *for* something. Such things may include human beings, non-human animals, organisms in general, species, or ecosystems, to take some standard candidates. This is closely tied to the question of what things can be harmed or benefited by a given change. Most will agree that human beings can be harmed or benefited, such that a geophysical change can be bad and possibly catastrophic if it causes physical suffering, psychological distress, economic cost, or some other form of hardship. Clearly climate change and other geophysical changes can be harmful in this way. Likewise, many non-human animals can experience pleasure or pain, making them susceptible to being harmed through suffering, at least. Matters become more controversial when we consider non-animal organisms, such as plants, fungi, and microorganisms. In some views, organisms of this kind can indeed be harmed and benefited, as they have good of their own (Goodpaster 1978; Taylor 1989). The same has been said of ecosystems as such (Callicott 1993), but that too is controversial. There is a similar controversy when it comes to the value of more abstract matters, such as biodiversity (Maier 2012). I will not assume these more controversial views, but I take it to be plausible and widely accepted that human beings and non-human animals can be harmed by geophysical changes. They are subjects that can undergo catastrophe.

The points just made are axiological in nature but not moral. So far I have made no claim about what has moral standing, i.e., what things deserve moral consideration. Just how far moral standing extends is reasonably debatable. Virtually all agree that human beings have moral standing. Some claim the same for non-human animals, organisms, species, and ecosystems (see Brennan and Lo 2024). In the interest of ecumenism, I will assume that human beings have moral standing while remaining agnostic about the moral standing of non-humans. Within the constraints of these self-imposed limitations, I hold that geophysical changes can be bad and potentially catastrophic in two broad ways. First, they can entail harm to human beings and non-human animals. Second, they can entail moral ills for human beings, by which I have in mind anything that is morally bad, such as distributive and procedural injustice, wrongful harm, and moral vice. The first is an axiological point, the second is a moral one. This distinction is needed. Although it is obvious that non-human animals can have experiences that are bad for them (i.e., suffering), it is not obvious that animal suffering is morally bad. It may be bad, as those who accept the moral standing of animals will hold, but I cannot assume that here. We need some way of differentiating between moral and non-moral badness, and the axiological/moral distinction allows for that. To give an example, a poultry farmer might acknowledge that his line of work is bad for chickens while also believing that there is nothing morally questionable about his line of work. We may disagree, but the farmer's view is perfectly intelligible and coherent.

Climate change, ocean acidification, mass extinction, nuclear war, and the rest are potentially catastrophic by virtue of the harms and moral ills they are likely to entail. In the following chapters, I defend various aspects of this value judgment.

Notes

1 As Oppenheimer et al. (2007) go on to argue, "With the general credibility of the science of climate change established, it is now equally important that policy-makers understand the more extreme possibilities that consensus may exclude or downplay."
2 Some material in this subsection was presented as "The Ethics of Geoengineering," Seventh Annual Conference of the International Society for Environmental Ethics, Allenspark, Colorado, June 2010.

References

Armstrong McKay, David I., Arie Staal, Jesse F. Abrams, Ricarda Winkelmann, Boris Sakschewski, Sina Loriani, Ingo Fetzer, Sarah E. Cornell, Johan Rockström, and Timothy M. Lenton. "Exceeding 1.5 C Global Warming Could Trigger Multiple Climate Tipping Points." *Science* 377, no. 6611 (2022): eabn7950.
Arrhenius, Svante. "Uber Den Einfluss Des Atmospharischen Kohlensauregehaltes Auf Die Temperatur Der Erdoberflache." *Bihang Till Kongl. Svenska Vetenskaps Akademiens Handlingar* 1, no. 22 (1896): 1–102.
Baum, Seth D. "Confronting the Threat of Nuclear Winter." *Futures* 72 (2015): 69–79.
Baum, Seth D., Timothy M. Maher Jr., and Jacob Haqq-Misra. "Double Catastrophe: Intermittent Stratospheric Geoengineering Induced by Societal Collapse." *Environment Systems & Decisions* 33, no. 1 (2013): 168–180.
Boettcher, Miranda, and Stefan Schäfer. "Reflecting Upon 10 Years of Geoengineering Research: Introduction to the Crutzen+ 10 Special Issue." *Earth's Future* 5, no. 3 (2017): 266–277.
Brands, Hal, and John Lewis Gaddis. "The New Cold War: America, China, and the Echoes of History." *Foreign Affairs* 100 (2021): 10.

Brennan, Andrew, and Norva Y. S. Lo. "Environmental Ethics", *The Stanford Encyclopedia of Philosophy* (Summer 2024 Edition), Edward N. Zalta & Uri Nodelman (eds.), 2024, https://plato.stanford.edu/archives/sum2024/entries/ethics-environmental/.

Broome, John. *Climate Matters: Ethics in a Warming World* (Norton Global Ethics Series). United States: W. W. Norton, 2012.

Callicott, J. Baird. "Toward a Global Environmental Ethic." *The Bucknell Review* 37, no. 2 (1993): 30.

Ceballos, Gerardo, Paul R. Ehrlich, and Peter H. Raven. "Vertebrates on the Brink as Indicators of Biological Annihilation and the Sixth Mass Extinction." *Proceedings of the National Academy of Sciences* 117, no. 24 (2020): 13596–13602.

Cornwall, Christopher E., Steeve Comeau, Niklas A. Kornder, Chris T. Perry, Ruben van Hooidonk, Thomas M. DeCarlo, Morgan S. Pratchett, et al. "Global Declines in Coral Reef Calcium Carbonate Production under Ocean Acidification and Warming." *Proceedings of the National Academy of Sciences* 118, no. 21 (2021): e2015265118.

Cowie, Robert H., Philippe Bouchet, and Benoît Fontaine. "The Sixth Mass Extinction: Fact, Fiction or Speculation?" *Biological Reviews* 97, no. 2 (2022): 640–663.

Crutzen, Paul J. "Albedo Enhancement by Stratospheric Sulfur Injections: A Contribution to Resolve a Policy Dilemma?" *Climatic Change* 77, no. 3–4 (2006): 211.

Davidson, Joe P. L., and Luke Kemp. "Climate Catastrophe: The Value of Envisioning the Worst-Case Scenarios of Climate Change." *Wiley Interdisciplinary Reviews: Climate Change* 15, no. 2 (2024): e871.

Dirzo, Rodolfo, Gerardo Ceballos, and Paul R. Ehrlich. "Circling the Drain: The Extinction Crisis and the Future of Humanity." *Philosophical Transactions of the Royal Society B* 377, no. 1857 (2022).

Ditlevsen, Peter, and Susanne Ditlevsen. "Warning of a Forthcoming Collapse of the Atlantic Meridional Overturning Circulation." *Nature Communications* 14, no. 1 (2023): 1–12.

Doney, Scott C., D. Shallin Busch, Sarah R. Cooley, and Kristy J. Kroeker. "The Impacts of Ocean Acidification on Marine Ecosystems and Reliant Human Communities." *Annual Review of Environment and Resources* 45 (2020): 83–112.

Doney, S. C., V. J. Fabry, R. A. Feely, and J. A. Kleypas. "Ocean Acidification: The Other CO2 Problem", *Annual Review of Marine Science* 1 (2009): 169–192.

Fleming, James Rodger. *Fixing the Sky: The Checkered History of Weather and Climate Control*. New York: Columbia University Press, 2010.

Gonçalves-Souza, Daniel, Peter H. Verburg, and Ricardo Dobrovolski. "Habitat Loss, Extinction Predictability and Conservation Efforts in the Terrestrial Ecoregions." *Biological Conservation* 246 (2020): 108579.

Goodpaster, Kenneth E. "On Being Morally Considerable." *The Journal of Philosophy* 75, no. 6 (1978): 308–325.

Harrison, Cheryl S., Tyler Rohr, Alice DuVivier, Elizabeth A. Maroon, Scott Bachman, Charles G. Bardeen, Joshua Coupe, et al. "A New Ocean State after Nuclear War." *AGU Advances* 3, no. 4 (2022): e2021AV000610.

Herzog, Stephen. "The Nuclear Fuel Cycle and the Proliferation 'Danger Zone'." *Journal for Peace and Nuclear Disarmament* 3, no. 1 (2020): 60–86.

Hoegh-Guldberg, O., P. J. Mumby, A. J. Hooten, R. S. Steneck, P. Greenfield, E. Gomez, C. D. Harvell, P. F. Sale, A. J. Edwards, K. Caldeira, N. Knowlton,

C. M. Eakin, R. Iglesias-Prieto, N. Muthiga, R. H. Bradbury, A. Dubi, and M. E. Hatziolos. "Coral Reefs Under Rapid Climate Change and Ocean Acidification." *Science* 318 (2007): 1737–1742.

Hogue, Aaron S., and Kathryn Breon. "The Greatest Threats to Species." *Conservation Science and Practice* 4, no. 5 (2022): e12670.

IMBIE Team. "Mass Balance of the Greenland Ice Sheet from 1992 to 2018." *Nature* 579 (2020): 233–239. https://doi.org/10.1038/s41586-019-1855-2.

IPCC. *Climate Change: The IPCC Impacts Assessment.* Contribution of Working Group II. [W.J. McG. Tegart, G.W. Sheldon, and D.C. Griffiths (eds.)]. Canberra, Australia: Australian Government Publishing Service, 1990.

IPCC. 2022: *Climate Change 2022: Impacts, Adaptation, and Vulnerability.* Contribution of Working Group II to the Sixth Assessment Report of the Intergovernmental Panel on Climate Change [H.-O. Pörtner, D. C. Roberts, M. Tignor, E. S. Poloczanska, K. Mintenbeck, A. Alegría, M. Craig, S. Langsdorf, S. Löschke, V. Möller, A. Okem, B. Rama (eds.)]. Cambridge, UK and New York, NY: Cambridge University Press, 2022, 3056 pp.

Jackson, L. C., R. Kahana, T. Graham, M. A. Ringer, T. Woollings, J. V. Mecking, and R. A. Wood. "Global and European Climate Impacts of a Slowdown of the AMOC in a High Resolution GCM." *Climate Dynamics* 45 (2015): 3299–3316.

Kaiho, Kunio. "Extinction Magnitude of Animals in the Near Future." *Scientific Reports* 12, no. 1 (2022): 19593.

Kaplow, Jeffrey M. "The Changing Face of Nuclear Proliferation." *International Studies Quarterly* 68, no. 2 (2024). https://doi.org/10.1093/isq/sqae045.

Keeling, Charles D. "The Concentration and Isotopic Abundances of Carbon Dioxide in Rural and Marine Air." *Geochimica et Cosmochimica Acta* 24, no. 3–4 (1961): 277–298.

Keith, David. "Geoengineering the Climate: History and Prospect." *Annual Review of Energy and the Environment*, 25 (2000), 245–284.

Keller, David P., Andrew Lenton, Emma W. Littleton, Andreas Oschlies, Vivian Scott, and Naomi E. Vaughan. "The Effects of Carbon Dioxide Removal on the Carbon Cycle." *Current Climate Change Reports* 4, no. 3 (2018): 250–265.

Kemp, Luke, Chi Xu, Joanna Depledge, Kristie L. Ebi, Goodwin Gibbins, Timothy A. Kohler, Johan Rockström, et al. "Climate Endgame: Exploring Catastrophic Climate Change Scenarios." *Proceedings of the National Academy of Sciences* 119, no. 34 (2022): e2108146119.

Leung, Jonathan Y. S., Sam Zhang, and Sean D. Connell. "Is Ocean Acidification Really a Threat to Marine Calcifiers? A Systematic Review and Meta-Analysis of 980+ Studies Spanning Two Decades." *Small* 18, no. 35 (2022): 2107407.

Liu, Jiajia, Ferry Slik, Shilu Zheng, and David B. Lindenmayer. "Undescribed Species Have Higher Extinction Risk Than Known Species." *Conservation Letters* 15, no. 3 (2022). https://doi.org/10.1111/conl.12876.

Maier, Donald S. *What's So Good about Biodiversity?* Dordrecht: Springer, 2012.

Matthews, H. Damon, and Ken Caldeira. "Transient Climate–Carbon Simulations of Planetary Geoengineering." *Proceedings of the National Academy of Sciences* 104, no. 24 (2007): 9949–9954.

Naughten, K. A., P. R. Holland, and J. De Rydt. "Unavoidable Future Increase in West Antarctic Ice-Shelf Melting over the Twenty-First Century." *Nature Climate Change* 13 (2023): 1222–1228.

Noble, T. L., E. J. Rohling, A. R. A. Aitken, H. C. Bostock, Zanna Chase, N. Gomez, L. M. Jong, et al. "The Sensitivity of the Antarctic Ice Sheet to a Changing Climate: Past, Present, and Future." *Reviews of Geophysics* 58, no. 4 (2020): 1944–9208.

Oppenheimer, Michael, Brian C. O'Neill, Mort Webster, and Shardul Agrawala. "The Limits of Consensus." *Science* 317, no. 5844 (2007): 1505–1506.

Orr, J. C., V. J. Fabry, O. Aumont, L. Bopp, S. C. Doney, R. A. Feely, A. Gnanadesikan, N. Gruber, A. Ishida, F. Joos, R. M. Key, K. Lindsay, E. Maier-Reimer, R. Matear, P. Monfray, A. Mouchet, R. G. Najjar, G.-K. Plattner, K. B. Rodgers, C. L. Sabine, J. L. Sarmiento, R. Schlitzer, R. D. Slater, I. J. Totterdell, M.-F. Weirig, Y. Yamanaka, and A. Yool. "Anthropogenic Ocean Acidification Over the Twenty-first Century and Its Impact on Calcifying Organisms", Nature 437 (2005): 681–686.

Pamplany, A., B. Gordijn, and P. Brereton. "The Ethics of Geoengineering: A Literature Review." *Science and Engineering Ethics* 26: (2020) 3069–3119.

Pan, Linda, Evelyn M. Powell, Konstantin Latychev, Jerry X. Mitrovica, Jessica R. Creveling, Natalya Gomez, Mark J. Hoggard, and Peter U. Clark. "Rapid Postglacial Rebound Amplifies Global Sea Level Rise Following West Antarctic Ice Sheet Collapse." *Science Advances* 7, no. 18 (2021). https://doi.org/10.1126/sciadv.abf7787.

Rahmstorf, Stefan, Jason E. Box, Georg Feulner, Michael E. Mann, Alexander Robinson, Scott Rutherford, and Erik J. Schaffernicht. "Exceptional Twentieth-Century Slowdown in Atlantic Ocean Overturning Circulation." *Nature Climate Change* 5, no. 5 (2015): 475–480.

Robock, Alan. "20 Reasons Why Geoengineering May Be a Bad Idea." *Bulletin of the Atomic Scientists* 64, no. 2 (2008): 14–18.

Robock, Alan. "Nuclear Winter." *Wiley Interdisciplinary Reviews: Climate Change* 1, no. 3 (2010): 418–427.

Robock, Alan, Luke Oman, and Georgiy L. Stenchikov. "Nuclear Winter Revisited with a Modern Climate Model and Current Nuclear Arsenals: Still Catastrophic Consequences." *Journal of Geophysical Research: Atmospheres* 112, no. D13 (2007).

Roe, Gerard H., and Marcia B. Baker. "Why Is Climate Sensitivity so Unpredictable?." *Science* 318, no. 5850 (2007): 629–632.

Román-Palacios, Cristian, and John J. Wiens. "Recent Responses to Climate Change Reveal the Drivers of Species Extinction and Survival." *Proceedings of the National Academy of Sciences* 117, no. 8 (2020): 4211–4217.

Rounsevell, Mark D. A., Mike Harfoot, Paula A. Harrison, Tim Newbold, Richard D. Gregory, and Georgina M. Mace. "A Biodiversity Target Based on Species Extinctions." *Science* 368, no. 6496 (2020): 1193–1195.

Sadai, Shaina, Alan Condron, Robert DeConto, and David Pollard. "Future Climate Response to Antarctic Ice Sheet Melt Caused by Anthropogenic Warming." *Science Advances* 6, no. 39 (2020): eaaz1169.

Sherwood, Steven C., Mark J. Webb, James D. Annan, Kyle C. Armour, Piers M. Forster, Julia C. Hargreaves, Gabriele Hegerl, et al. "An Assessment of Earth's Climate Sensitivity Using Multiple Lines of Evidence." *Reviews of Geophysics* 58, no. 4 (2020): 93.

Smith, Jordan P., John A. Dykema, and David W. Keith. "Production of Sulfates Onboard an Aircraft: Implications for the Cost and Feasibility of Stratospheric Solar Geoengineering." *Earth and Space Science* 5, no. 4 (2018): 150–162.

Sokolski, Henry D., ed. *Getting MAD: Nuclear Mutual Assured Destruction, Its Origins and Practice.* Strategic Studies Institute, US Army War College, 2004.

Svoboda, Toby. *The Ethics of Climate Engineering.* New York: Routledge, 2017.

Svoboda, Toby, Klaus Keller, Marlos Goes, and Nancy Tuana. "Sulfate Aerosol Geoengineering: The Question of Justice." *Public Affairs Quarterly* 25, no. 3 (2011): 157–179.

Taylor, Paul W. *Respect for Nature: A Theory of Environmental Ethics*. Princeton: Princeton University Press, 1989.
Tertrais, Bruno. "'On the Brink'—Really? Revisiting Nuclear Close Calls Since 1945." *The Washington Quarterly* 40, no. 2 (2017): 51–66.
Weitzman, Martin L. "On Modeling and Interpreting the Economics of Catastrophic Climate Change." *The Review of Economics and Statistics* 91, no. 1 (2009): 1–19.
Westing, Arthur H. "The Ecological Dimension of Nuclear War." *Environmental Conservation* 14, no. 4 (1987): 295–306.

2 Understanding Pessimism

Ecological pessimism is simply the expectation that the future will involve ecological catastrophe of some kind. This definition is vague in at least two ways. First, there is vagueness in what counts as an expectation. One might take that to vary from a faint suspicion all the way up to absolute certainty. Second, there is vagueness in what counts as an ecological catastrophe. The potential variance there is also great, ranging from moderate climate change to the extinction of all life on earth. It is appropriate to leave the definition vague, however. The term "ecological pessimism" merely refers to a general state of mind or, more specifically, a kind of attitude. It can take many different forms, as we shall see. For this reason, I will not offer a precise definition of ecological pessimism as such. The vague definition will suffice. Instead, I will try to be precise when it comes to outlining my own version of ecological pessimism, as well as the reasons that support it. In short, there are many ways to be an ecological pessimist, so it will not be useful to precisify the general concept. Nonetheless, there are some general concerns about this vagueness that I should address before moving to more substantive matters.

Vague Expectations and Vague Catastrophes

One might think that the two types of vagueness just noted are problematic. Imagine someone who takes herself to be an ecological pessimist, because she has a suspicion that the earth will warm by 1°C relative to the pre-industrial average temperature. In my view, this is very optimistic! If my definition of ecological pessimism permits this kind of view, then one might claim that there is something wrong with the vague definition I have offered. But this is just a standard problem of vagueness, which accompanies many different concepts that we are nonetheless able to employ in useful ways. Consider the concept of tall. If one likes, he could claim that a person measuring five feet in height is tall. Most of us would not agree with that, but we probably would allow a person measuring seven feet in

DOI: 10.4324/9781003570523-3

height to be tall. The more interesting cases are in the middle. Is a person measuring 5 feet and 10 inches in height tall? Here the vagueness in the concept of tall might pose a problem, but in the other two cases, things are reasonably clear. The five-foot person is not tall, and the seven-foot person is tall, and we can be confident of this even if we don't know exactly where to draw the line between tall and not tall.

Something similar goes for ecological pessimism. We might reasonably disagree about where to draw the line separating genuine ecological catastrophe from ecological change that is not catastrophic. Although perhaps interesting for scholars working on the question of vagueness itself, drawing that line seems to me useless and unnecessary for the purposes of the arguments I offer in this book. It is reasonably clear that some ecological changes are catastrophic: mass extinction, boiling oceans, widespread eco-systemic collapse, global warming of 8°C, a runaway greenhouse effect, and so on. Just as almost no one reasonably doubts that the seven-foot person is tall, almost no one can reasonably doubt that these outcomes would be catastrophic in an ecological sense. Personally, I do not think that one degree of warming would count as catastrophic, but my view on that hardly matters, because I believe that we are on course for much worse.

This ties into the other vague item in our definition, namely what counts as an expectation. I offer the same account here as in the preceding paragraph. Does an uninformed hunch count as an expectation? I do not know, but it does not matter here. The evidence supporting events that plausibly count as ecologically catastrophic is strong. It falls short of certainty, of course, but the available information, including projections about the future, is grim. This pertains to both the human and non-human worlds, so to speak. The latter includes geophysical and biological phenomena, while the former includes social phenomena. Put briefly, non-human nature as we know it is in bad shape and getting worse, and human societies are not doing nearly enough to address that problem. Very roughly, my case for ecological pessimism rests on two observations: first, if we do not respond quickly and seriously to various ecological crises, catastrophe is likely; second, our response has been and likely will remain, neither quick nor serious. Whatever vagueness attaches to the concept of expectation, it does not plausibly threaten my position. My expectation of ecological catastrophe is not a whim or suspicion but rather a reasonable attitude based on evidence from the natural and social sciences. At least, I will try to show that this is the case. The reader might end up disagreeing, but it is doubtful that any perceived problems with my view will have to do with the vagueness I have noted here.

Pessimism in Philosophy

Pessimism has been rare in philosophy, both historically and at the present time. It is often seen as an attitude to be resisted (see Moellendorf 2024). Among major figures, Schopenhauer offers the only obvious case. As Bertrand Russell notes, aside from Schopenhauer, virtually all other Western philosophers have been optimists in some sense (Russell 2004). This has come in many varieties of moral philosophy, metaphysics, epistemology, and political thought: faith in divine salvation, Kant's postulates of immortality and freedom, utilitarians' hope for social progress, certain Marxists' expectation of the revolution, the conviction that this is the best of all possible worlds, the belief that God ensures human knowledge, and more. One might argue that skeptics should count as pessimists, given their doubts about knowledge, whether globally or in some specific domain. Perhaps so, but skeptics have always constituted a small minority. The majority of philosophers seem to take skepticism as a challenge, working to show that knowledge, or something like it, is possible. If philosophers generally treat skepticism as something to be overcome, then that attitude alone is a kind of optimism, for it holds out hope for the possibility of success.

To be clear, I am not arguing against optimism in philosophy, nor am I making the charge that optimistic philosophers are naive. Perhaps it is entirely appropriate for philosophers to be optimistic. After all, if everyone was a skeptic, for example, there might be little left for philosophers to do. I shall leave it to others to determine whether or not that would be a bad thing. On the other hand, we might at least wonder whether this reveals a certain bias. If there is a general bias among philosophers (and others) in favor of optimism, then we should be on guard against dismissing pessimistic positions too easily. Whatever the reason, it is striking that so few Western philosophers have been pessimists.

Unfortunately, Schopenhauer's philosophy will not offer much help for my case. His is a metaphysical pessimism, rooted in some rather implausible claims about the "Will" as the thing in itself. Roughly put, everything in the world is striving to achieve some end but is doomed to failure. One sees this in the case of human desire, which is just one manifestation of the Will. I desire something. That desire is either satisfied or not. If it is satisfied, the desire is soon replaced by a new desire, which in turn is either satisfied or not. There is no end to this cycle, aside from the state of boredom, which is also a kind of dissatisfaction. It is not possible to achieve any lasting fulfillment of my desires, because there is no stable, lasting state of satisfaction. New desires will always intrude, or I will languish in boredom. For Schopenhauer, this is not just a contingent fact about human beings, but rather the fundamental nature of all things. His philosophy is very interesting, and I have written about it elsewhere (Svoboda 2022), but it will not serve as anything like a foundation for my arguments in this book.[1] Unlike Schopenhauer, I am not a philosophical pessimist in general, nor do I subscribe to any sort of metaphysical pessimism. Rather, mine is an empirical pessimism, having to do with facts about both the natural world and human societies. Some of these facts may well be contingent. I do not think they are expressions of some necessary, fundamental principle. Presumably, humanity could do much more to reduce the probability of catastrophe. We just do not care to do so.

The Nature of the Pessimistic Attitude

I have described ecological pessimism as a state of mind but more specifically as an attitude. What type of attitude is involved here? It is common to distinguish between cognitive and non-cognitive attitudes. The former includes beliefs, while the latter includes desires and emotions. A standard account of the difference relies on the notion of "direction of fit." Beliefs are mental states with a "mind-to-world" direction of fit, while desires are mental states with a "world-to-mind" direction of fit. This is to say that beliefs aim to represent the world, while desires aim to change the world. When I believe something, I am purporting to match my mind to the way the world really is. For instance, I might believe that it is raining outside because I observe that vehicles are using windshield wipers. This mental state is either true or false, depending on whether it is in fact raining. Conversely, when I desire something, I want the world to fit with my mental state. For instance, if I am thirsty, I might desire that there is a water fountain in front of me, even if there is no realistic chance of this being the case in present circumstances. On the standard picture, desires are neither true nor false. Because they do not purport to represent the way the world is, they cannot be faulted for failing to match the world. Their very nature is to demand that the world be in a certain state, and this can include states that are very different from the actual one. I will rely on this fairly uncontroversial account of beliefs and desires as typifying, respectively, cognitive and non-cognitive attitudes.

As I shall use the term, "ecological pessimism" refers to a cognitive attitude, although it is likely to be accompanied by various non-cognitive attitudes as well. The expectation of ecological catastrophe is, at its core, a belief that such catastrophe is likely to occur in the future. This belief might cause, or otherwise be connected to, non-cognitive attitudes like anxiety, fear, sadness, dread, and the like, but I do not take any of these desire-like attitudes to be the core of the ecological pessimism I defend. One might question this. Why should, say, dread with regard to impending climate change not count as ecological pessimism? To clarify, and in accordance with the vague definition offered above, I think it is perfectly reasonable to classify such an attitude as being ecologically pessimistic.

This is just not the attitude I am interested in understanding and defending in this book. Rather, my focus is a cognitive attitude. Again, one might question this. Why focus on this cognitive attitude rather than one of the non-cognitive attitudes that could count as ecological pessimism?

A cognitive analysis opens up a rich, epistemological terrain. It is usually granted that non-cognitive attitudes are not truth-apt, which is to say that they are not properly designated as true or false. This is not surprising, as such attitudes do not purport to fit with the world or to tell us the way things are. On the other hand, cognitive attitudes do exactly that. They are therefore truth-apt. If I believe that it is raining outside, my belief is true or false, and this depends on the way the world actually is. Beliefs can be evaluated as true or false, justified or unjustified, and probable or improbable. Desires do not lend themselves to such epistemological analysis, at least not so easily. Some philosophers have tried to develop such accounts, with Simon Blackburn's quasi-realism being perhaps the best-known example, but they remain controversial (Blackburn 1993). Can the feeling of dread regarding future climate change be classified as true? Perhaps in a minimalist sense (Smith 1994). I do not take a position here on whether quasi-realist accounts are likely to be successful. However, there is no similar controversy surrounding the cognitivist approach, which can help itself to a fully robust set of epistemological evaluations. Because of this, there is much to say about a cognitivist variety of ecological pessimism. We can ask whether it is true and evaluate the degree to which it is justified. It is not clear that we can do this with a non-cognitivist variety of ecological pessimism.

Perhaps most importantly, a cognitivist conception of ecological pessimism allows us to embed it within chains of reasoning. As I said, the core of this attitude consists of a belief that ecological catastrophe is likely. Like any proposition, "that ecological catastrophe is likely" can be plugged into standard forms of logical argument, such as modus ponens. Due to the so-called Frege-Geach problem, it is not clear that non-cognitive judgments can be embedded within such forms of logical argument. If the relevant attitudes are not truth-apt, then how could we apply logical operations to them, such as negation? Again, Blackburn has developed a "logic of attitudes" that is meant to address this problem. Perhaps something like that can work, but this is a matter of some controversy. While non-cognitive ecological pessimism would be of psychological interest and maybe more, a cognitivist conception of such pessimism unquestionably allows for logical embedding. This renders it a matter of philosophical interest.

Moral Pessimism

As I will argue throughout this book, there are many reasons why it is reasonable to expect ecological catastrophe in the future. Elsewhere I have defended moral pessimism, which we may define here as the expectation that, in general, human beings will not act morally (Svoboda 2022). I will not reproduce the full argument here, but I can outline the general idea behind it. It is beyond question that human history has been full of atrocities, including such lovely phenomena as genocide, aggressive war, terrorism, and many types of oppression. As a matter of common sense, it is obvious that these things have been very bad in a moral sense. Although certain varieties of moral ill may recede from time to time, they are soon enough replaced by others, making it difficult to discern anything like moral progress in our history. For instance, one society might cease practicing chattel slavery but move on to threatening civilian populations with nuclear annihilation. If we specifically look to our present time, there is little cause for optimism. Although the injustices and harms of our own time are not identical to those of, say, a century ago, it is certainly not obvious that we are better off morally. We still find aggressive war, if in new forms, and we now pose a non-negligible risk to the survival of civilization itself. If anything, matters are worse now, given our increased capacity for carrying out atrocities. Because of all this, it is reasonable not to expect humanity to comply with their moral obligations. We shall return to this matter throughout the book.

In practical terms, the moral pessimist will not be surprised when human beings, either collectively or individually, behave immorally. This is to be expected, given our history. Such behavior can range from minor wrongdoings to great injustices. Of course, one might be committed to moral pessimism in a theoretical sense but struggle to live according to it in a practical sense. This is not uncommon when it comes to philosophical ideas. Those who are sympathetic to radical forms of skepticism often find it difficult to suspend belief when it comes to daily life, for example (Burnyeat 1983). This does not mean that moral pessimism is false. The difficulty of adopting it as a practical attitude may owe to certain social or psychological issues. On the other hand, some individuals might succeed when it comes to incorporating that attitude into their lives. I return to the question of how the ecological pessimist should live in later chapters.

It is important to distinguish between what is to be expected and what is obligatory. The moral pessimist is not a person with lax moral standards but rather a person who does not expect humanity to abide by those standards, whatever they are. In everyday discourse, these two issues are often conflated. When pointing out some moral problem, one often hears the response, "What else did you expect?" Of course, the mere fact that some sort of behavior is to be expected does not excuse it. Typically, dictators engage in brutal suppression of their populations, but that does not change the fact that such suppression is reprehensible. I expect dictators to carry on being reprehensible in the future, but that expectation has nothing to do with fixing appropriate moral standards for judging their behavior. As we shall see, the same goes for the ecological pessimist, who expects humanity to bring about catastrophe in the future and judges that behavior to be abhorrent. That expectation does not alter the relevant moral standards.

Moral pessimism is distinct from ecological pessimism. It is clearly possible to expect that human beings will fail to comply with their moral obligations in general (moral pessimism) without expecting ecological catastrophe to occur. There are many ways for human beings to be morally abhorrent, many of which do not necessarily involve ecological catastrophe. Likewise, it is possible to expect ecological catastrophe (ecological pessimism) without expecting human beings to fail in their moral obligations. For instance, one might think that catastrophe is likely due to some natural event over which humans have no influence, such as an asteroid impact. All of this is to say that neither type of pessimism entails the other, whether in a logical or a psychological sense.

At the same time, these two pessimisms are complementary in our present context. The ecological crises we face are humanity's doing, and I will make the case that they carry grave moral ills. To take one example, anthropogenic climate change involves severe injustice. John Broome helpfully makes the case for this, noting seven relevant features of the greenhouse gas emissions driving climate change (Broome 2012, 55–59). First, they cause serious harm, including death and injury due to severe weather events. Second, that harm is the result of something we do rather than something we merely allow. Consistent with common sense and many moral theories, causing harm is usually thought to be worse than merely letting harm occur, all else being equal. Third, the harm caused by our emissions is not accidental. Sometimes accidental harm can be excused or responsibility for it mitigated, such as when that harm was not reasonably foreseeable, but this is not so with our emissions. Fourth, we do almost nothing to compensate the victims of climate change. Fifth, we engage in most emitting activities for our own benefit, such as the pleasure of traveling.[2] Sixth, the harm of emissions does not come close to being fully reciprocated. Rich individuals and countries emit greenhouse gasses at very high rates compared to poor individuals and countries, and yet the latter group often experiences the worst harms of climate change. Broome contrasts this with the case of traffic congestion. The presence of other vehicles on the road harms me by slowing my commute and introducing various annoyances, but I am contributing to that problem just as much as other motorists. I cannot claim to be a victim of injustice here. We might say that I am simply paying the price of helping myself to the benefits of using the available infrastructure. Seventh and finally, we could reduce our emissions substantially without much cost to ourselves: increasing energy efficiency, reducing our consumption, or shifting to the use of renewable forms of energy. We have made little progress in this area. Broome's account here does not exhaust every way in which our emissions might be unjust, and perhaps one can quibble with certain points, but on the whole, this provides a fairly convincing set of reasons as to why climate change is *prima facie* unjust.

As this quick account suggests, and we shall see in greater detail in subsequent chapters, the environmental damage being done by humanity is plausibly viewed as a moral issue. We are not like some invasive species of a plant that devastates an ecosystem but which cannot be properly held to moral account. We are conscious, rational beings with enough flexibility in our behavior that, unlike the invasive plant, we need not devastate our ecologies. Imagine, contrary to fact, that something about our physiology necessitated ecological catastrophe, say the secretion of some radioactive isotope that poisoned other species. Arguably, we would not be morally blameworthy for the devastation that we caused. To be sure, that devastation would remain harmful, but it would be a great misfortune or tragedy rather than a matter of moral culpability. Here one could be an ecological pessimist without being a moral pessimist. It would not be our moral badness driving ecological catastrophe.

Of course, in the actual world, we cannot avail ourselves of such excuses, at least not while remaining reasonable. We need not destroy the rest of nature, but we are doing so anyway. That says something horrifying about us, I presume. For my part, I subscribe to both ecological and moral pessimism. Although the two are not necessarily connected, in our present case, they are contingently connected. I will make the case throughout this book that, as a matter of fact, our ecological prospects are tied to our moral prospects. In many cases, it is the same behavior (e.g., inaction on climate change) that is both morally bad and ecologically dangerous. If we suddenly became serious about climate justice, that would reduce the probability of climate-related catastrophe. Unfortunately, there is little reason to be optimistic on that front.

The foregoing might invite a certain misunderstanding, so I need to clarify. In saying that moral and ecological pessimism are contingently connected, I do not mean to say that one explains the other. In particular, I do not claim to know why humans behave as they do. In my book defending a certain type of misanthropy, I stressed that my project was descriptive rather than explanatory (Svoboda 2022). The central claim there was that humanity is morally bad, as supported by a great deal of historical and contemporary evidence. Importantly, I did not claim that humanity's moral badness *causes* its reprehensible behavior. Instead, I claimed that, given humanity's reprehensible behavior, the species is rightly judged to be morally bad. The best explanation for why humanity is the way it is would presumably depend on various empirical factors, including matters of psychology and social conditions. Again, the misanthropic view I defended was descriptive in nature. I did not pretend to offer any causal analysis.

Something similar holds for the connection of moral and ecological pessimism. I do not claim in the present book that our moral badness is the cause of the ecological catastrophe that we can plausibly expect, nor do I claim that certain insights of moral pessimism can be used to explain why we are headed toward an unseemly ecological future. To be fair, I do not deny those claims either. Nonetheless, whatever the causal entanglements might be, there is clearly a correlation between our moral badness and the ecological ills we are currently causing. Even as a purely descriptive matter, this is interesting for various reasons. For one thing, it suggests that we need not choose between averting moral catastrophe and being moral. Returning to the example of climate change, reducing our greenhouse gas emissions would reduce the risk of both climate catastrophe and climate injustice, all else being equal. Making progress in one area can reasonably be expected also to help in the other. Conversely, failing to make progress in one area can reasonably be expected also to hinder progress in the other. The latter scenario aptly describes our current state of affairs.

This moral pessimism harmonizes with certain varieties of misanthropy, a topic that has received some attention in recent years (Gerber 2021; Kidd 2021; Norlock 2021; Svoboda 2022). If we are pessimistic about the moral prospects of humanity, then it is reasonable to maintain a negative assessment of humans in general. There is no question that human beings have engaged in morally monstrous behavior, and there is little reason to think that will stop. Such behavior can be seen in our treatment of non-human nature, particularly animals (Cooper 2018). Nonetheless, we should be careful to distinguish between pessimism and misanthropy, which are often conflated (Kidd 2022). As David Cooper points out, pessimism concerns "aspects ... that are destructive of the possibility of happiness," whereas misanthropy has to do with "failings ... for which humankind is answerable and rightly held to account" (Cooper 2018, 4–5). In short, the pessimist expects things to be bad, the misanthrope thinks humans are bad, and these two judgments may or may not be connected. It is possible to be a pessimist, even a moral pessimist, without thereby condemning humanity. For instance, one might think that human nature makes certain moral flaws tragically unavoidable. Such a person might then feel compassion for the species, or at any rate, abstain from holding it accountable for its failings. Whether that stance would be defensible is a further question I do not address here. Nonetheless, this possibility shows that, although the ecological pessimism I sketch here is suggestive of certain forms of misanthropy, it does not require it.

About What Should We Be Pessimistic?

A pessimistic attitude will be distinctively ecological by virtue of the objects of that attitude. Roughly put, if one is pessimistic about matters ecological, then one is ecologically pessimistic. Of course, satisfying that condition does not automatically make one an ecological pessimist of the variety studied in this book. One might be pessimistic about the recovery prospects of some small, local ecosystems, for example. This person is not thereby committed to the expectation that ecological *catastrophe* is likely, for she might in general be optimistic about the planet's prospects. Someone who has the *attitude* of ecological pessimism will be pessimistic about matters ecological, but this will not be limited to a few cases. Rather, the genuine ecological pessimist expects very bad things on a global scale. In this section, I aim to specify some of the objects, the targets, of that attitude.

Clearly, any pessimistic attitude will involve some kind of judgment of badness. In the case of ecological pessimism, we judge that our ecological prospects are bleak and that likely future scenarios will be bad in some way. I have already addressed what general moral ills are likely to come in the future, including harm and injustice, but this does not yet address the following question: about what, specifically, should we be pessimistic? This is a distinct question because any harm or injustice must be harm or injustice to some subject, whether human or non-human. There are many possible victims in the catastrophic scenarios that might occur in the future. Of course, with the exception of the very near future, we cannot identify precisely which individuals will be harmed or subjected to injustice in the future. This is because we have no way of knowing, or even plausibly conjecturing, which individual humans or non-humans will happen to exist, say, a thousand years from now. Nonetheless, we can identify likely phenomena that, should they come to pass, will be very bad, in a morally relevant sense, for some parties, whoever they turn out to be.

Such phenomena plausibly include the harm experienced by sentient entities, both human and non-human. Due to climate change, human beings will suffer from a wide range of phenomena in the future, including food insecurity, displacement, and severe weather events such as intense heat waves. This harm will come in the form of physical pain, psychological distress, and economic damage. Because such harm is likely to occur, and because it will be driven by ecological phenomena, it is an appropriate target of our ecological pessimism. Similarly, climate change will cause great harm to non-human animals (Nolt 2011b) via precipitation change and other factors. Climate change is the most obvious driver, but we may add to this the many other ecological harms accruing to non-humans, such as ocean acidification and habitat loss through human development. Although non-human sentient entities cannot experience economic damages, many of them can suffer, and this is reasonably taken to be a bad thing, regardless of one's stance on whether we have moral obligations to non-human animals. I will return to this point later in the course of directly addressing objections to my view. For now, let me observe that harm to non-human animals is ordinarily viewed as a bad thing, often in some moral sense, say lighting a cat on fire (Harman 1977). One need not believe in anything like animal rights in order to assent to this view. One might argue that such harm is justified, or perhaps that its badness is not of a moral variety, but it is not controversial to say that the harm itself (e.g., the cat's suffering) is bad.

Aside from suffering, we might also look forward to the many deaths promised by our ecological crises. John Nolt estimates that the average American's lifetime greenhouse gas emissions will be causally responsible for the deaths of "one or two" future human beings (Nolt 2011a). Of course, the estimate is rough, making various assumptions and relying on projections that are themselves uncertain, but Nolt is conservative in his estimate, so the true number of deaths is probably higher. Regardless, the point stands that many of us are causal contributors to an ecological phenomenon that will kill many people in the future, hundreds of millions if we rely on Nolt's numbers. That seems bad. Likewise, although I am not aware of an analogous estimate, it is no doubt the case that many non-human organisms will die in the course of our ongoing and future crises. Matters become even worse when we remember that climate change is only one of our crises and that other phenomena will kill additional humans and non-humans. To anticipate an objection, we cannot deflect the foregoing simply by noting that these casualties would have occurred in any case. Although it is true that human and non-human organisms are certain to die eventually, the victims of climate change will typically be premature deaths. Dying from climate change is not bad simply because it confirms one's mortality but rather because that death deprives one of future life they might otherwise have enjoyed. Once again, I will not say much here in response to potential objections. Instead, I will address objections directly in a later chapter.[3]

Next, let us consider the vast landscape of ecological injustice. It is clear that environmental hazards tend to be unequally distributed along racial and economic lines, with poor and minority communities taking the brunt of those burdens (Mohai et al. 2009; Shrader-Frechette 2002). *Prima facie*, this is a form of environmental injustice. Further, I am aware of no theory of distributive justice that would permit such a distribution. Egalitarians may object to the obvious inequality in how environmental burdens are shared. It is true that most egalitarian theories allow for unequal distributions under the right conditions, most famously Rawls' Difference Principle, which allows for social and economic inequalities in some cases.

However, a necessary condition of justice under the Difference Principle is that those inequalities must benefit those who are the worse off (Rawls 1971). That is virtually never the case, of course. Typically, it is the worse-off members of society who shoulder the burdens (e.g., pollution), while the relatively well-off enjoy the benefits (e.g., cheap energy) of the activities that produce those burdens. Likewise, desert-based theories of justice are unlikely to permit the distribution of environmental benefits and burdens that we often observe, as there is no plausible case to be made that members of poor and minority communities deserve to shoulder these burdens. Unfortunately, there is little reason to think that such injustice will be limited to our own time.

I will mention one more example of a fitting target for our ecological pessimism, namely the disappearance of much that is beautiful and interesting in non-human nature. Humanity is causing a mass extinction event and destroying natural wonders, such as coral reef ecosystems. Obviously, this is regrettable, even if the destruction is not in itself morally bad. I do not take a position here on whether we have genuinely moral reasons for preserving species or beautiful landscapes, but our failure to do so is unfortunate, even if in a purely aesthetic sense. We are producing a world that will be less biologically diverse, more barren, and less ecologically interesting than it has been in the past. This seems an appropriate target for pessimism.

I have mentioned mass extinction. One very real possibility is the extinction of the human species. This could plausibly occur as a result of any of several ecological catastrophes that leave the planet uninhabitable for our kind: extreme climate change, the collapse of ecosystems due to mass extinction, nuclear or biological warfare, as well as others. Would that be a bad thing? Obviously, the misery presumably experienced by human beings in the course of an extinction event would be bad, but that is not the issue here. The question is whether the fact of our extinction would be a bad thing in itself. Nearly every species that has ever existed on earth has disappeared. We typically do not think it a tragedy that the various species of trilobites are no longer with us. Why should we judge differently when ours joins that list? We like to think of our own species as being especially important, perhaps uniquely valuable. Presumably, some ancient trilobite would have felt the same about its own species, had it the capacity to do so. Of course, this invites the response that, because humans do have certain capacities that other species lack, the extinction of ours would be especially or even uniquely bad. Such capacities might include our alleged rationality, the scope of our imagination (including the appreciation of past and future), our wide range for desire and valuation, the ability to produce beauty, and so on. Should our species disappear, the universe would lose all of those good things going forward.

Let us allow that the disappearance of humanity would indeed carry these costs. The question remains: Would humanity's extinction be bad all things considered? Perhaps not. Although we do have capacities for creating value, such as art and meaningful relationships, we also have capacities for creating a great deal of disvalue. Indeed, one and the same capacity can be used to bring about either value or disvalue, depending on choice and circumstances. In a way, this echoes Kant's observation that the only unconditional good thing is goodwill, as any other candidate can become bad under certain conditions (Kant 1996). Even putative virtues can be put to ill use. Courage is normally a good thing, but not when it is used to carry out a terrorist attack. Likewise, our intelligence as a species has produced both intellectually valuable discoveries and weapons of mass destruction. In attempting to determine whether humanity's extinction would be all things considered bad, we need to consider both the advantages and disadvantages it would bring. I agree that it would be a shame to lose the future goods that humanity otherwise might have created, but it would be a relief for the world to be freed of the many ills humanity has always brought about: war, genocide, the destruction of non-human nature, racism, the mass suffering of sentient entities for the sake of cheap food, and many others. On balance then, it is not clear that humanity's extinction would be a bad thing on the whole. Elsewhere I have made the argument that the extinction of our species would in fact be a good thing (Svoboda 2022), but I will not assume that here. By questioning whether human extinction would be bad, I am depriving myself of a source of support for my view, as I cannot now claim that our extinction is something about which we should be pessimistic. If anything, that prospect is a cause for optimism. Accordingly, if I am mistaken about that, and human extinction actually is a bad thing, then that just adds another item to the list of potential bad things we can expect in the course of future catastrophes.

Notes

1　For a very useful account of Schopenhauer's pessimism, see Janaway (1999).
2　Though it is difficult to conceive, apparently some people enjoy this activity.
3　Whether or not death is harmful to the person who dies is a classic philosophical puzzle that I will not claim to solve here. However, even if one holds that death is not a harm, it is difficult to deny that death is usually bad in some sense.

References

Blackburn, Simon. *Essays in Quasi-Realism*. New York: Oxford University Press, 1993.
Broome, John. *Climate Matters: Ethics in a Warming World*. New York: W. W. Norton, 2012.

Burnyeat, M. F. "Can the Skeptic Live His Skepticism?" In *The Skeptical Tradition*, edited by M. F. Burnyeat. Berkeley: University of California Press, 1983, 117–148.

Cooper, David E. *Animals and Misanthropy*. New York: Routledge, 2018.

Gerber, Lisa. "A Word against Misanthropy." *Journal of Philosophical Research* 46 (2021): 73–87.

Harman, Gilbert. *The Nature of Morality*. New York: Oxford University Press, 1977.

Janaway, Christopher. "Schopenhauer's Pessimism," *Royal Institute of Philosophy Supplements* 44 (1999): 47–63.

Kant, Immanuel. *Groundwork for the Metaphysics of Morals*. In *Practical Philosophy*, translated and edited by Mary J. Gregor. New York: Cambridge University Press, 1996.

Kidd, Ian James. "Varieties of Philosophical Misanthropy." *Journal of Philosophical Research* 46 (2021): 27–44.

Kidd, Ian James. "Misanthropy and the Hatred of Humanity." In *The Moral Psychology of Hate*, edited by Noell Birondo. Lanham: Rowman and Littlefield, 2022.

Mohai, Paul, David Pellow, and J. Timmons Roberts. "Environmental Justice." *Annual Review of Environment and Resources* 34 (2009): 405–430.

Moellendorf, Darrel. "Mobilizing Hope against Pessimism and Plutocracy." *Ethics, Policy & Environment* 27 (2024): 1–17.

Nolt, John. "How Harmful Are the Average American's Greenhouse Gas Emissions?" *Ethics, Policy and Environment* 14 (2011a): 3–10.

Nolt, John. "Nonanthropocentric Climate Ethics." *Wiley Interdisciplinary Reviews: Climate Change* 2, no. 5 (2011b): 701–711.

Norlock, Kathryn J. "Misanthropy and Misanthropes." *Journal of Philosophical Research* 46 (2021): 45–58.

Rawls, John. *A Theory of Justice*. Harvard: Harvard University Press, 1971.

Russell, Bertrand. *History of Western Philosophy*. New York: Routledge, 2004.

Shrader-Frechette, Kristin. *Environmental Justice: Creating Equality, Reclaiming Democracy*. New York: Oxford University Press, 2002.

Smith, Michael. "Why Expressivists about Value Should Love Minimalism about Truth." *Analysis* 54, no. 1 (1994): 1–11.

Svoboda, Toby. *A Philosophical Defense of Misanthropy*. New York: Routledge, 2022.

3 Evil in Environmental Affairs

There is much that is evil in human affairs. Some of it takes an environmental form. This fact gives us yet another reason to adopt ecological pessimism.

Defining Evil

As with many moral concepts, the search for necessary and sufficient conditions of evil seems hopeless.[1] There have been numerous attempts, and although many suggested conditions seem initially plausible, all of them are susceptible to plausible counter-examples.[2] At least for my purposes, a set of necessary and sufficient conditions for evil is not needed. I will instead rely on a descriptive account of evil, suggesting certain conditions under which an action ordinarily counts as evil. Those conditions might not be sufficient. For instance, there may be some action that is normally evil but might be non-evil in extreme emergencies. Likewise, my conditions might not be necessary, as there may be other ways of engaging in evil action that I have not considered. Because of this, accepting my account requires tolerating some degree of vagueness. This is a price worth paying, however, as it allows me to avoid the acute problem of counter-examples, which is faced by virtually all theories that seek to provide necessary and sufficient conditions of evil. Of course, my account might still be susceptible to counter-examples in a way, say because some cases do not fit with the general description I provide, but that will need to be considered case by case. My aim is only to offer a plausible account of what generally counts as evil, admit the possibility of occasional exceptions to that account, and use it to argue that certain human actions plausibly qualify as evil. Assuming it is possible, a set of necessary and sufficient conditions would be philosophically preferable, but given that various philosophers have searched for this and not found it, a more modest approach is reasonable.

Arguably, there is less need to delineate exactly what counts as evil than there is to delineate other moral concepts. Despite various differences,

DOI: 10.4324/9781003570523-4

virtually all commentators agree that the term "evil" is to be reserved for the most atrocious of actions. As it happens, such actions are not very subtle and thus are likely to be noticeable for those who care to look. Leaving aside positions taken in bad faith or for self-serving political reasons, cases of evil are often likely to be fairly obvious due to their magnitude. We do not need a finely honed, precise theory of evil in order to know that mass killing of the innocent is evil. Contrast this with other moral concepts, such as responsibility, obligation, blame, and so on. By its nature, the charge of evil is less subtle than other moral charges. It is supposed to be a blunt concept, applied in extreme cases. If this is true, then leaving our account of evil action somewhat vague might not be problematic, even if similar vagueness for other moral concepts would be problematic.

Further, I opt for an ecumenical approach to understanding evil action. This is appropriate, because my aim is not to make a contribution to the philosophical understanding of counts as evil, but rather to argue that certain activities within "environmental affairs" are evil. Despite their important theoretical differences, many competing theories of evil converge when it comes to judging certain actions to be evil, although each theory may provide a very different account as to why any particular action is evil. For instance, any plausible theory of evil action needs to cover genocide, even though the "evil-making" features of genocide are open to reasonable debate. Perhaps genocide is evil because of the quantity of harm it causes, the fact that it seeks to exterminate a particular identity, that it undermines the basic functioning of persons, or something else. We need not settle on a single theory in order to be fairly confident that genocide is evil. This is an attractive feature of my ecumenical approach, as my argument that some action is evil will not depend on whether some particular, sure-to-be-controversial account is ultimately correct.

This raises a question: Why discuss theoretical accounts of evil at all? I have already allowed that genocide is obviously evil if anything is, so what use is there in examining evil in a theoretical way? Why not simply rely on intuition to identify evil actions? The answer is that, in some cases, evil is not immediately obvious. We might agree that genocide in general is evil, but we might not recognize that some particular case of genocide is evil, say by conveniently denying that the case qualifies as genocide. Humans are exceptionally skilled at rationalization, especially when that is reinforced within a group with similar interests. Thus we might overlook some evil for psychological, economic, or social reasons. It is difficult to find an atrocity in human history whose agents admitted the truth of what they were doing, rather than attempting to justify their actions to themselves and others in various ways, including moral appeals. This can obscure the truth, even for those who are not operating in bad faith.

Relying on some theoretical approach, even an ecumenical one, can provide some protection against this, provided that one is honest and fair in their judgment. If we pursue that course, we might discover that, on reflection, some previously overlooked action is evil. A purely intuitive approach might pick out obvious cases, but it would not suffice for non-obvious cases. That is why theoretical tools will be useful to some degree. An example of such a case is provided by climate obstructionism, which I discuss below. At first glance, it might appear unreasonable to accuse climate obstructionists of evil, but I believe that charge becomes very plausible after some theoretical reflection.

Is Evil Qualitatively Distinct from Other Moral Concepts?

In the literature on evil, one of the main issues is determining whether evil is qualitatively distinct from other moral concepts, such as wrongness. If yes, then we must specify what quality is distinctive of evil actions. Most philosophers who write on the subject think that evil is indeed qualitatively distinct. Russell is a prominent exception, arguing that evil is only quantitatively distinct from ordinary wrongdoing. On one variety of this quantitative view, "evil actions are simply wrong actions that are extremely harmful, regardless of the psychology of the evildoer, and regardless of the kinds of harm inflicted" (Russell 2007, 676; see also Russell 2014). As Russell notes, this account has the virtue of simplicity. A more complex yet still "psychologically thin" version might

> define evil actions as culpably wrong actions that have a certain kind of connection to extreme harms, in that they either produce extreme harms, are intended to produce extreme harms, contribute to extreme harms, or are acts of appreciation of extreme harm.
>
> (Russell 2007, 676)

In both accounts, evil action is determined by the severity of the harms involved. The latter is notable in that it allows us to make sense of collective evil actions, for it treats contributions to extreme harms as evil. This is attractive, because some apparently clear cases of evil in history involved contributions from many agents.

As for those who believe that there is something qualitatively distinctive about evil action, many competing accounts have been offered. For example, perhaps evil actions cause substantial harms that ruin lives (Formosa 2013; 2019), perhaps they involve "intolerable" harms, such as denying persons their basic needs (Card 2002), or perhaps an evil action is one that interferes with a person's fundamental agency (Kekes 1998). The challenge for any such account is to specify plausible criteria of evil action while preserving evils actions' qualitative distinctiveness. Russell argues that some attempts ultimately collapse into quantitative accounts that treat evil action as very harmful or especially wrong (Russell 2007). For example, we might agree that interfering with someone's fundamental agency is evil but only because it is very harmful to do so.

In order to resist that, a proponent of qualitative distinctiveness must explain why some feature of an action is evil-making, and they must do so in a way that is not simply reducible to the quantity of harm connected to that action. It is not always clear that such proponents are successful. One reason for this is that for any suggested criteria for evil action, it is plausible to see the qualifying actions as very harmful, especially bad, or very wrong in some other sense. Paradigm cases of evil (e.g., genocide) do often involve a great quantity of harm, so we might be skeptical that any further quality is needed in order for the relevant actions to count as evil.

The focus of this chapter is on evil action, rather than evil intentions, motivations, or character. It may be that evil actions are, sometimes or always, connected to certain dimensions of the agent, but I will remain agnostic regarding that issue. There is a rich terrain for psychological inquiry here, both in the empirical and philosophical senses, but they are not immediately important for the purposes of this book. We are examining the—in my view likely catastrophic—ecological impacts of human action. Certain of these actions, as I will argue, plausibly count as evil. However, I shall not seek to explain why human beings perform such actions, a matter that might ultimately have something to do with intention, motivation, character, or the like. Whatever the reason, humanity has performed actions that count as evil.

The sort of evil I have in mind is *morally* disvaluable. This is to distinguish it from what is sometimes called (e.g., in the philosophy of religion) "natural evil," such as the destruction and suffering caused by an earthquake. What precisely counts as a morally disvaluable phenomenon is contentious, with competing theories offering different views. Nonetheless, we can identify some cases that virtually everyone will agree are morally, and not just naturally, bad: murder, genocide, betrayal of friends for personal gain, and so on.

Of course, not all morally bad phenomena are reasonably described as evil. Telling a lie in order to avoid embarrassment may be wrong, but it is not evil. In addition to being morally disvaluable, then, genuinely evil phenomena must be "bad enough." The term is to be reserved for phenomena that are especially bad. Plausibly, this includes mass extermination, terrorism, genocide, and the like. What makes something especially bad may be merely quantitative, such as crossing some threshold of harm, or it might be qualitative in nature, as noted above.

What actions may we consider especially bad in a moral sense? In my view, an action or a set of actions is typically evil if (but not only if) (1) it causes extensive, unjust harm to a large group of persons and (2) this effect is reasonably foreseeable by the acting party. There may be other sufficient criteria for evil as well, but the foregoing are attractive because they seem plausible in their own right. These criteria also correctly identify and explain evident cases. For instance, it is fairly obvious that any case of genocide will be extensively and unjustly harmful to a large group of people, and of course its perpetrators will foresee this effect, as the purpose of genocide is to exterminate a group of people.

There is vagueness here, of course. At what point has some foreseeable, unjust, harmful action crossed the threshold into the realm of evil? As with many accounts and concepts, it is difficult to draw a non-arbitrary line here regarding what does or does not meet the relevant criteria. This is not a problem for my account, however, because it is only an epistemological problem when it comes to identifying borderline cases. Some cases will clearly surpass any threshold, despite their vagueness: genocide, targeted killings of civilians, mass torture, destruction of population centers, and so on.

My account remains ecumenical in that it is open to many possible explanations as to why some act is evil, including explanations generated by other theories of evil. My claim is not that an action is evil *because* it foreseeably causes unjust harm, but rather that actions that foreseeably cause unjust harm will typically be evil. Perhaps this is due to the fact that such actions also tend to destroy lives, deny basic needs, or undermine fundamental aspects of agency. It certainly seems plausible to believe that extensive, foreseeable, unjust harm will usually carry with it these other deleterious aspects. Again, I offer no opinion on the ultimate, fundamental features of evil. What I say below regarding the evil of climate obstruction-ism and other actions is compatible with many different theories of evil.

Finally, I agree with Russell that it is desirable for an account of evil action to remain psychologically thin, thereby averting the baggage that comes with tying the moral evaluation of action to the mental states of agents. My account is psychologically thin, because it references the mental states of neither agents nor victims of evil. All that is needed is extensive, unjust harm that is reasonably foreseeable. Importantly, this harm need not be foreseen by its perpetrators. Through willful ignorance or simple disinterest, an agent might not consider the unjust harm that their action will likely cause. Nonetheless, if that unjust and extensive harm was reasonably *foreseeable*, then the action still counts as evil on my account. The actual psychology of the agent does not matter.

By contrast, psychologically thick accounts of evil action appeal to the mental states of agents and/or victims of evil, but they face both substan-tive and epistemic problems. First, it is a complicated matter to specify just what mental states are relevant to evil action. These could include intentions, motivations, and vices, for example. Second, diagnosing actual cases of evil will often be difficult if we need to determine the mental states of the agents involved, as these are often obscured, intentionally or otherwise. A psychologically thin account can simply avoid these problems. Of course, these considerations are far too brief to warrant dismissing psychologically thick accounts, but they do indicate that a thin account has certain advantages. Let us now look at an example of evil action.

Why Climate Obstructionism Is Evil

To some, it might appear hyperbolic to declare climate obstructionism to be evil, but the claim is very plausible.

By "climate obstructionism," I mean any coordinated attempt to hinder, slow, or undermine progress in addressing the crisis of climate change. Such obstructionism can take many forms: denying the reality that the climate is changing, manufacturing doubt regarding the anthropogenic nature of climate change, downplaying the severity or probability of potential impacts, lying about the costs of renewable energy, harassing scientists, and many more. I say that climate obstructionism must be coordinated in order to distinguish it from the attitudes, beliefs, or actions of individuals, which by themselves are unlikely to obstruct climate progress. Genuine climate obstructionism, as I shall use the term, requires coordination among various actors and will often take political forms.[3]

By "progress in addressing the crisis of climate change," I mean any coordinated effort to mitigate climate change or its impacts. This too can take many forms: international agreements among states to limit their greenhouse gas emissions, financial incentives such as a carbon tax, domestic laws regulating fossil fuels, social policies to pursue adaptation to changing environments, the use of geoengineering technologies to avert some of the expected impacts of climate change, and so on. Leave aside debates about which of these efforts is to be preferred. That is an important question, of course, but not relevant here. As with climate obstructionism, I understand climate progress to require more than individual actions or resolutions. Given the global nature of the problem, individual action is unlikely to be efficacious. Genuine progress requires some degree of social coordination.

There is no doubt that climate obstructionism has been, and continues to be, a reality in our world, especially in the United States. Examples are numerous, many of which are well known. There is the case of Exxon Mobil hiding the findings of its own scientists regarding the dangers posed by greenhouse gas emissions, campaigns of disinformation meant to sow unreasonable doubt among the public about the reality and impacts of climate change, explicit opposition to meaningful climate policies on the part of the fossil fuel industry, and of course the refusal of various politicians, especially within from members of the extremist Republican party, to pass serious climate legislation at the state and federal levels (Matthews and Eaton 2023; Oreskes and Conway 2010, 2011).

The question I address here is how we should regard this obstructionism in a moral sense. There are many possible answers. We might judge it to be morally praiseworthy, permissible, impermissible, regrettable, unjust to a minor degree, or something else. In my view, climate obstructionism is morally reprehensible to the point of being genuinely evil. My argument for this will come in two parts. First, I sketch what plausibly counts as evil in the relevant sense. Second, I show that climate obstructionism fits with this standard and thus qualifies as evil.

I will now make the case that climate obstructionism is evil. Briefly put, climate obstructionism (1) causes extensive, unjust harm to a large group of people and (2) this is reasonably foreseeable by the obstructionists themselves.

Regarding (1), it is scientifically well-established that climate change is occurring, that it is caused primarily by anthropogenic emissions of greenhouse gasses, and that it will cause substantial harm to many people. This harm comes in many varieties: sea-level rise, extreme weather events, the spread of disease to new areas, food insecurity, and more (IPCC 2022). There is no doubt that climate change will harm many individuals, especially when we take into account the future generations that will be forced to deal with the impacts. Finally, climate change carries a non-negligible risk of human extinction.

Moreover, ethicists widely agree that many of these harms are unjust (see Broome 2012). One need not be a trained ethicist to see the injustice. Those who are likely to experience the worst impacts of climate change are low-emitters in the global South, i.e., parties that have near-zero responsibility for climate change and who have the fewest resources to adapt to it. It is intuitively clear that this extensive harm is unjust. If nothing else, we have a situation in which the wealthy countries of the world are imposing very substantial harms on the inhabitants of poor countries, including death (Nolt 2011).

Climate obstructionism helps cause this unjust outcome. By its very nature, such obstructionism aims to allow the unjust harms of climate change to occur unabated. Further, it is clear that climate obstructionists have had a great deal of success, especially in the United States, where serious action on addressing the crisis has been undermined and delayed. If not for climate obstructionists impeding progress, it is likely that humanity would have pursued more aggressive mitigation of emissions and adaptation to changing climatic conditions. Both of these courses would reduce the harm of climate change, probably to an extensive degree.

As for (2), there is no question that the effects of climate change are reasonably foreseeable. Most obviously, many obstructionists *do* foresee the effects. Many of them, especially politicians, political appointees, lobbyists, and think-tank researchers are highly educated individuals with access to the relevant information, yet they proceed with obstructing climate progress. Additionally, the relevant science has been clear for several decades, so they cannot reasonably claim that the effects of their actions are not foreseeable.

Objections

Let us consider some objections to my account.

Mistaken Beliefs and Misguided Actions Are Not Evil

First, it might be objected that climate obstructionism is not genuinely evil, at least in many cases, because it depends upon mistaken beliefs and the misguided actions that result therefrom. Some people honestly think that climate change is a hoax, for example. Arguably because of this, they do not consider progress in fighting climate change to be a priority. They then support politicians and policies that ignore and worsen the climate crisis. All of this may be regrettable, misguided, foolish, and perhaps even carry some degree of moral culpability, but surely it is not evil to hold and act upon mistaken beliefs.

This offers an opportunity to clarify something. I do not consider the person described above to be a climate obstructionist. The genuine obstructionist is the agent who manipulates the person above into holding those false beliefs and supporting misguided policies. In other words, I do not have in mind the average voter or motorist, but rather those who know perfectly well what they are doing. The vast majority of powerful individuals (e.g., politicians and fossil fuel lobbyists) who have undermined climate progress are educated and well aware that climate change is real and threatens massive harm. Nonetheless, and usually for personal gain, they choose to lie, distract, and confuse others when it comes to climate change. Such agents will not be saved by the objection, because they are not plausibly viewed as dupes who have simply made an honest mistake in belief.

Charges of Evil Are Politically Dangerous

One might object that talk of evil, especially accusations against individuals, is politically dangerous. For example, such talk might encourage fanaticism or excessively punitive measures. This could lead to a situation in which the virtuous "climate saints" feel justified in persecuting "climate heretics," possibly harming the innocent in the course of pursuing their righteous cause. It is not surprising that charges of evil are commonly issued by religious fanatics, who, when they have power, cause a great deal of harm and injustice. We should not encourage climate activists to take a similar stance.

As an initial response, I submit that it is simply true that climate obstructionism is evil. The truth value of that claim does not depend on the alleged fact that it encourages immoral or antisocial behavior. Setting that aside, however, it seems clear that judgments of evil are compatible with remaining rational and non-fanatical. Many of us judge certain events of the twentieth century to be morally evil, and yet few seem fanatical in the way they do so. Moreover, any cause, no matter how noble, is susceptible to being captured by fanatics who find themselves with power. This might happen in the case of climate activism, but presumably, we have moral reasons to prevent that.

Charges of Evil Are Superfluous

Next, it might be charged that talk of evil is superfluous and unnecessary, adding nothing more to our moral judgments, except perhaps some rhetorical force. That is, we might judge some cases to be unjustly and extensively harmful in a way that was reasonably foreseeable by the agent and simply stop there. This objection offers a challenge: What value is there in talk of evil, aside perhaps from noting that some phenomenon is especially bad (see Calder 2013)?

I claim that moral evil is conceptually, and not merely rhetorically, useful. This is because it captures the thought that certain actions are morally bad in a distinct way. Again, both apartheid and lying to an acquaintance are morally bad, but they are not bad in the same way. There is something about apartheid that requires a different moral category than what is needed for everyday lies. The concept of evil accommodates the thought that apartheid is morally bad in a deep, serious fashion. It thereby allows us to avoid the implausible view that humanity's worst crimes are no different in kind from minor moral offenses.

Nietzsche's and Scheler's Challenge

Another, more general line of attack is to adopt a sort of Nietzschean perspective, treating attributions of evil as unbecoming of a noble person, as they are *ressentiment*-laden accusations (see Card 2002, 27–49). The concern here is that charges of evil are made in a kind of bad faith and from a position of weakness, concealing the accuser's own inadequacy and self-loathing. Perhaps we are just envious of those with power, using morality as a vehicle to vent our impotent hatred of them. This strikes me as an important counter-perspective to my own view, so I will address it in detail, specifically by examining the work of the now-neglected philosopher, Max Scheler.

Scheler's *Ressentiment* is a response to Nietzsche's critique of Christian morality as laid out in *The Genealogy of Morals*. Nietzsche claims that Christianity is a hatred-inspired inversion of values in which weakness displaces strength. In his view, the Christian values humility, charity, and equality only because he is too weak to attain higher goods like strength, nobility, and power. Christian morality is therefore typified by *ressentiment*, which is a denigration of that which one secretly desires to be but is incapable of becoming. Scheler counters that "true" Christianity is not motivated by *ressentiment*. Nietzsche philosophically misjudges the character of Christian morality and historically mistakes certain decadent moralities for genuine Christianity. Although Scheler admits that Christianity is in some ways liable to being perverted by *ressentiment*, there is no reason to suppose that it *must* be thus perverted. On the contrary, Nietzsche should have directed his critique against "modern humanitarianism," which privileges utility values above "higher" spiritual values. Scheler thinks this modern morality is responsible for the "decadence" that Nietzsche so rightly deplores. Scheler commits considerable energy to level the Nietzschean critique against this modern morality, portraying "love of man," "altruism," and movements like socialism and communism as being fueled by *ressentiment*. Although Scheler is right to reject Nietzsche's claim that Christian morality is necessarily motivated by *ressentiment*, Scheler himself makes the same mistake vis-à-vis modern morality. One of Scheler's objections to Nietzsche can also be deployed against his own argument, because although modern morality *may* be especially prone to *ressentiment*, there is no ground for supposing any necessary connection between the two. Scheler offers no compelling reason to believe that practitioners of modern morality are always motivated by *ressentiment*. Just as the Christian is liable to *ressentiment* but need not succumb to it, so with the modern humanitarian. Scheler only outlines possible opportunities for *ressentiment* to arise within modern morality—a valuable project, but not one sufficient to prove his thesis. While it may be true that modern humanitarianism fails to appreciate the highest values, it would be wrong to see this failure as necessarily inspired by *ressentiment*. One aim of this paper is to illustrate the philosophical value of engaging with Scheler, who has mostly been overlooked in contemporary philosophy (although cf. Poellner 2022).[4]

Nietzsche's critique of Christian morality finds its most focused expression in *The Genealogy of Morals*, aptly subtitled "*Eine Streitschrift*," "A Polemic." This book introduces the notion of *ressentiment*, which Scheler says is the "most profound" discovery made about the "origin of moral judgments" in recent philosophy (Scheler 1994, 27). Nietzsche's "discovery" is that moral judgments can be motivated by a secret hatred that seeks to reverse the natural hierarchy of values in order to exalt weakness and denigrate strength. While certain moralities may appear altruistic, mild, or compassionate, they can actually be infused with a desire to tear down the noble few who have set themselves apart from the "herd." Nietzsche refers to this as the "will to power of the weak," because it is carried out by feeble people whose aim is "*poisoning the consciences* of the fortunate with their own misery, with all misery, so that one day the fortunate [begin] to be ashamed of their good fortune...." These are the "men of *ressentiment*," who thirst for revenge against those who are noble and strong, carrying out their vengeance under the guise of morality (Nietzsche 2000). Their goal is to pervert the natural order of values, making the strong appear evil and the weak appear good. This is the sole manner in which the weak can gain some measure of power, since they have no hope of achieving prominence within the unpolluted ethical order that values precisely what they are incapable of becoming (see Reginster 1997).

Nietzsche identifies Christian morality as a primary vehicle of *ressentiment*, sarcastically quipping that Dante should have inscribed above the gateway to paradise, "I too was created by eternal *hate*" (Nietzsche 2000, 1.15). For Nietzsche, the Christian lives in a sort of bad faith, deceptively claiming that weakness, impotence, "anxious lowliness," and subordination are merit, "goodness of heart," humility, and obedience, respectively. Such a "miserable" person is incapable of revenge, but prefers to call his impotence "forgiveness," and he christens his cowardice as "patience." All these deceptive equivocations come from a "workshop where *ideals are manufactured*" (Nietzsche 2000, 1.14). In Nietzsche's view, Christian morality exalts humility, obedience, long-suffering, and forgiveness not because it originally esteems these characteristics as virtues, but rather because Christianity wishes to have weapons in its battle against the strong and noble. Unable to acquire the characteristics she would otherwise admire, the Christian makes a virtue out of necessity, blessing her weakness as goodness and venomously denouncing the virtues of the strong. This two-fold movement—(1) reconceiving one's weakness as strength and (2) denigrating the strengths of others as bad or evil—is the very essence of *ressentiment*, and Nietzsche sees it at the heart of Christian morality.

While Scheler strongly rejects the association of *ressentiment* with Christianity, he accepts Nietzsche's diagnosis that *ressentiment* can and does poison moral judgments. Scheler agrees with Nietzsche in seeing *ressentiment* as "a self-poisoning of the mind" that produces "a tendency to indulge in certain kinds of value delusions," which are prompted by emotions like "revenge, hatred, malice, envy, the impulse to detract, and spite" (Scheler 1994, 29). Moreover, "*Ressentiment* can only arise if these emotions are particularly powerful and yet must be suppressed because they are coupled with the feeling that one cannot act them out—either because of weakness, physical or mental, or because of fear" (Scheler 1994, 31). It is entirely different with the "noble person," who has a "non-reflective awareness of his own value and of his fullness of being" (Scheler 1994, 37). Such a noble person never questions his own worth, but a weak person is perpetually unsure of his own value and thus feels the need to bolster himself by attacking the noble. Although the weak person claims to value his own weakness, he originally does so only as a means of dealing with his inferiority to the noble person. In fact, *ressentiment* can be formulated in this manner: "A is affirmed, valued, and praised not for its own intrinsic quality, but with the unverbalized intention of denying, devaluating, and denigrating B. A is "played off' against B" (Scheler 1994, 49).

In the end, however, the bearer of *ressentiment* becomes her own victim, because "the values themselves are inverted." Whereas one initially affirms A only because it will denigrate B, this reversal eventually becomes permanent and genuine, i.e., the poison infects the natural order of values itself. One no longer merely pretends to despise B and affirm A, but "those values which are positive to any normal feeling become negative" (Scheler 1994, 56). At first the slavish person detracts from the noble person but now that which is slavish in oneself detracts from that which is noble in oneself. This is the attitude Nietzsche attacks in his critique of asceticism, which he sees as a turning against the self in a spirit of hatred and revenge. Persons of *ressentiment* are not so much dishonest as they are sick. Scheler and Nietzsche are thus very close in their conception of *ressentiment*, viewing it as an attitude that begins with hatred of the noble other and ends in hatred of the self.

The interesting split between Nietzsche and Scheler occurs when the latter refuses to agree that Christian morality is guilty of *ressentiment*. Scheler thinks Nietzsche makes this mistaken diagnosis for two reasons. First, Nietzsche *philosophically* misjudges the essence of Christian morality. Scheler claims that Nietzsche judges Christian morality according to a false standard, failing to take into account that Christianity is a religion whose highest values are not the biological and vital ones that Nietzsche favors. Scheler notes that Christianity locates the highest good—the "kingdom of God"—beyond this world, so it is only appropriate that its accompanying morality should also transcend the "sphere of life." If one lacks this understanding, then Christian morality certainly will appear decadent, but this will not appear to be the case when one remembers that Christianity is not primarily interested in cultivating biological values. This, according to Scheler, is precisely Nietzsche's mistake, and herein lies his philosophical misjudgment (Scheler 1994, 82–83). Christian morality fails to grant biological values the highest place, not because it is tinged with a secret hatred of those goods which it cannot hope to possess, but rather because it genuinely esteems certain spiritual values as higher than the vital ones. If Scheler is right, then Christian morality is not subject to the formulation of *ressentiment* laid out above. A is not affirmed for the sake of degrading B, but rather A is affirmed because it is originally recognized as a higher value than B. In failing to see this, Scheler says, Nietzsche mistakenly treats Christian morality as a mere vehicle of *ressentiment*.

In the second reason for his error, Nietzsche mistakes certain *historical* deformations of Christianity for the "true" Christianity. Nietzsche confuses "genuine" Christian morality with such modern movements as socialism and communism. According to Scheler, however, Christianity has little in common with these egalitarian movements, as it is "aristocratic" and has never affirmed the "equality of souls before God" (Scheler 1994, 86). Further, genuine Christian morality does not entail the decadence and sickness that Nietzsche claims. Far from commanding its adherents to *eliminate* hostility, subjugation, and power in order to become docile and insensitive, its "virtue lies in the free sacrifice of these impulses, and of the actions expressing them, in favor of the more valuable acts of "forgiveness' and "toleration'" (Scheler 1994, 86). These natural impulses belong to the "complete living being," and any morality that demands their *cessation* is indeed decadent and life-denying, i.e. deserving of Nietzsche's critique. Christian morality, however, makes no such demand. According to Scheler, it only urges its practitioners to subordinate these life impulses to the spiritual values that are expressed in forgiveness and the like. Hence if a particular morality includes decadent, life-denying imperative, it is not genuinely Christian, even if it calls itself so. For instance, "Christian" socialism bears nothing in common with genuine Christian love of neighbor, because the former is not inspired by the spirituality of the latter, socialism being nothing more than a degenerate form of "modern humanitarianism" Failing to recognize this, Nietzsche wrongly identifies the two and mistakenly believes that Christianity is historically typified by *ressentiment*.

Scheler does offer an interesting aside that is meant to further explain Nietzsche's mistake:

> Nietzsche is right in pointing out that the *priest* is most exposed to this danger [of *ressentiment*], though the conclusions about religious morality which he draws from this insight are inadmissible. It is true that the very requirements of his profession... expose the priest more than any other human type to the creeping poison of *ressentiment*.
>
> (Scheler 1994, 46–47)

The "requirements of his profession" include the fact that the priest is not supported by secular power and that he *must* control emotions of "revenge, wrath, hatred." Such requirements do indeed make the priest liable to *ressentiment*, since lack of secular power and repression of emotion can produce the weakness and desire for revenge that are the essential features of *ressentiment*. Nonetheless, Nietzsche is wrong to conclude from this that the priest *must* be motivated by *ressentiment*, because there is no *necessary* connection between that moral perversion and the "requirements of his profession." One's being subject to a certain danger is not a sufficient condition for one's necessarily falling victim to that danger. Hence, as Scheler says, Nietzsche's correct observation does not justify his "conclusions about religious morality," i.e., that the priest and his kind are in fact subject to *ressentiment*. It is entirely possible that the priest should skirt or overcome the "dangers of his profession." Elsewhere, Scheler notes that "Christian values can very easily be perverted into *ressentiment* values" but that the "core" of Christianity is not tainted with *ressentiment* (Scheler 1994, 61). Scheler is quite right on this point. One can imagine situations in which Christian morality is used as a weapon of revenge. For example, a poor person might "forgive" a rich person for being economically better off, but the poor person may only be using this "forgiveness" as a tool by which to feel self-righteous, secretly hating the rich person all along in his attempt to bolster his own self-esteem. Nonetheless, one can also imagine situations in which forgiveness is pure and motivated by honest compassion, and Nietzsche has provided no argument against this very real possibility.

Although Scheler vindicates Christian morality in this manner, he immediately aims the same Nietzschean critique against "modern humanitarianism," which contrary to Christianity "does not command and value the *personal* act of love from man to man, but primarily the impersonal '*institution*' of welfare'" (Scheler 1994, 97). This is the essential difference between the two: Christianity values the personal love of neighbor, modern morality values the abstract well-being of "man." Moreover, Christianity locates the highest values "beyond" this world, i.e., its highest values are spiritual. Modern humanitarianism, on the other hand, treats earthly welfare as the highest good, having no use for spiritual values. Scheler thinks this movement, often associated with Christianity in only a superficial way, is fueled by *ressentiment* and finds its expression in the attempt to redistribute wealth, erase political and socio-economic distinctions between persons, remove national boundaries, and destroy all values that issue from "knighthood" and the "caste of warriors" (Scheler 1994, 96–97). This humanitarianism is a "*ressentiment* phenomenon" because

> this socio-historical emotion is by no means based on a spontaneous and original *affirmation of a positive value*, but on a *protest, a counter*

impulse (hatred, envy, revenge, etc.) against ruling minorities that are known to be in possession of positive values.

(Scheler, 1994, 98)

Humanitarians call for political and socio-economic equality not because they positively value equality, but rather because they themselves are not capable of attaining the "positive values" of the minority of superior persons. Being inferior in value, the weak seek to level the playing field by dragging down their betters, all the while disguising their hatred and desire for revenge with moral rhetoric. Hence modern humanitarianism, for Scheler, fits the template of *ressentiment*. Scheler's account should be of interest to scholars working on *ressentiment* in general and not just to those interested in Nietzsche's use of the idea (see Hoggett 2018; Meltzer and Musolf 2002; Morelli 1998; TenHouten 2018).

Although he offers numerous speculations about why humanitarianism is *ressentiment*-laden, Scheler never succeeds in providing a concentrated argument that demonstrates this. If one wishes to establish the actual truth of Scheler's thesis, one must do more than simply *imagine* ways in which modern morality might have sprung from *ressentiment*. Many of Scheler's complaints against modern humanitarianism—e.g., its ignoring spiritual values, its destruction of differences in status among persons, its replacing personal love of neighbor with impersonal love of mankind—*may* be appropriate for other reasons, but they do not entail the *ressentiment* of modern morality. In fact, one of Scheler's objections to Nietzsche's critique of Christian morality applies to Scheler's own critique of modern morality. Having admitted that Christian morality is in some ways especially prone to *ressentiment*, Scheler notes that the "core" of Christianity remains untainted and that *ressentiment* is only a perversion. For example, although the priest is particularly subject to *ressentiment*, he can and often does overcome the temptation. In deeming modern morality to be always and everywhere motivated by *ressentiment*, Scheler forgets that practitioners of a morality that is liable to *ressentiment* need not fall victim to it. Just as the priest may resist the lure of *ressentiment*, so may the modern humanitarian.

Scheler identifies a number of reasons why modern morality is a *ressentiment* phenomenon. For example, it purports to love the human *species* rather than human persons. Further, modern morality disdains the "immediate circle of the community and its inherent values," seeing every community as equal (Scheler 1994, 99). Scheler holds that these traits qualify modern humanitarianism as a *ressentiment* phenomenon. One loves the whole species because he hates the higher individual specimens, and one refuses to recognize a hierarchy of communities for the same reason. This is motivated by a secret hatred, a desire for revenge. But Scheler does not provide a convincing argument that this is so. It is not clear why the characteristics of modern morality necessarily entail *ressentiment*. At best, Scheler merely illustrates the possibility that humanitarianism could be tainted with *ressentiment*, but illustrating a possibility and demonstrating a necessary connection are quite different. Love for the species *might* be produced by a hatred of higher specimens, but it might also be produced by a sincere compassion for the species. The former case would be *ressentiment*, but the latter would be a genuine and positive valuation. It very well could be the case that the modern humanitarian loves the species and bears no ill will toward those whom Scheler deems superior. Since both are very real possibilities, one cannot point to love of the species as evidence of *ressentiment*, since there is no established connection between the two. One would have to make recourse to further evidence, but Scheler does not take this route, simply assuming that love of the species entails a hatred of higher persons. As an empirical matter of fact, it may often be true that the modern humanitarian labors under *ressentiment*, but there is no reason to believe that the modern humanitarian *must* labor under *ressentiment*. The same holds true for those who reject "the immediate circle of the community." Scheler supposes that this is done only when one harbors a secret hatred for higher communities, seeking to revenge herself by destroying all differences and recognizing every community as equal. This is, of course, possible, but again Scheler does not show why this must be so. One can easily imagine a person who honestly believes that all communities are equal, who positively values this equality, and who bears no grudge against any particular sort of community. Whether right or wrong in this belief, this person would not be guilty of *ressentiment*. (see Kelley 2016).

It is true that Scheler believes in an objective hierarchy of values that he calls the *ordo amoris*, which is phenomenologically given and affectively known. As Scheler says in "*Ordo Amoris*," "The heart possesses a strict analogue of logic in its own domain." This logic is "an *ordre du coeur*, a *logique du coeur*, a *mathematique du coeur* as rigorous, as objective, as absolute, and as inviolable as the propositions and inferences of deductive logic" (Scheler 1973, 117). In this order of preferencing, spiritual values hold a higher place than utility and biological values. If Scheler is correct about the *ordo amoris*, then modern humanitarianism is highly deficient, because it ignores spiritual values and accords utility values a primary place. Nonetheless, this inversion of the objective hierarchy alone would be no evidence of *ressentiment* in modern morality, since there are other potential causes that could be responsible for the deficiency. To prove his thesis, Scheler would have to show that modern morality's inversion of the objective order is caused by a desire for revenge against the bearers of the higher values. Further, he must also show that this revenge is motivated by an inability on the part of modern humanitarians to realize those higher values. Unfortunately, Scheler provides no argument that purports to demonstrate this.

It is clear that Scheler disapproves of modern morality, finding it deficient in honoring utility values to the detriment of spiritual and personalist values. He may be right in seeing modern morality as incomplete in this regard, but this criticism is different from his criticism that modern morality is motivated by *ressentiment*. Nonetheless, Scheler often conflates the two, implying that the former entails the latter. For instance, it may be true that modern love for the species is inferior to the personal love of Christianity, but contrary to Scheler this need not have anything to do with *ressentiment*. As discussed above, one could sincerely value love of the species in its own right, perhaps honestly overlooking the importance of personal love. In this case, one's order of values may be deficient, but there is no trace of *ressentiment* to be found. To take another example, one might earnestly believe that spiritual values are based on nonsense, preferring instead to follow a morality that seeks to better the lives of persons on earth. If Scheler is right, this person's morality is lacking an important spiritual aspect, but there is no reason to think that it is moved by *ressentiment*. One can be honestly mistaken without harboring a secret hatred or desire for revenge against others. Since it is obvious that there can be such cases, Scheler's claim that modern humanitarianism is always motivated by *ressentiment* goes much too far.

Scheler's book offers many useful insights, particularly his refutation of Nietzsche's claim that Christianity is *ressentiment*-laden. Christian morality may be prone to *ressentiment*, but there is no reason to suppose that its practitioners must fall victim to that moral perversion. In fact, any "Christian" morality that is fueled by *ressentiment* does not deserve to be called Christianity at all. As discussed above, Scheler is quite right on this point. Unfortunately, Scheler proceeds to make the same mistake as Nietzsche, the only difference being that the target is modern morality rather than Christianity. One can refute Scheler's claim that modern humanitarianism is a *ressentiment* phenomenon by using the same argument he uses against Nietzsche. Although modern morality is in many ways liable to the poison of *ressentiment*, the former does not entail the latter, because it is quite possible that humanitarians should be sincere and free of hatred. Modern morality's proneness to *ressentiment* is no more a mark against it than it is for Christianity. Despite this mistake, however, Scheler's diagnosis of humanitarianism remains valuable. Although one cannot accept the claim that modern morality is thoroughly infected with *ressentiment*, one can recognize that it is liable to such an infection, and Scheler's thought provides tools to protect against this possibility. Just as Nietzsche did Christianity a service by discovering potential pitfalls in

Christian morality, so does Scheler serve modern morality. The Christian need not fully accept Nietzsche's diagnosis, and she can use Nietzsche's insights to identify possible shortcomings in her own moral practice. Likewise, the modern humanitarian need not agree completely with Scheler in order to use his thought as a guard against the creeping potentiality of *ressentiment*. For this reason, despite the fact that one cannot accept Scheler's primary thesis, *Ressentiment* remains an important contribution to the study of ethics.

Why Evil Invites Ecological Pessimism

There is a plausible connection between evil in environmental affairs and ecological pessimism. Evil actions of the sort we have reviewed reliably involve catastrophe and in two senses. First, they cause catastrophic harm. The victims of evil actions suffer extensive harm in the form of widespread death, injury, and deprivation of various goods (e.g., social, cultural, and economic goods). Second, evil actions are morally catastrophic in their own right. Evil actions are by their nature especially bad in a moral sense, covering the most atrocious acts committed by human beings. We might say that the agent of evil has failed catastrophically as a moral agent. I will rely on both senses of catastrophe below.

Although not a necessity, it is likely that our human future will resemble our human past. This holds especially for the relatively near-term future. If we look at human history, both recent and more distant, we find a pattern of evil actions: genocide, slavery, mass killing, terrorism, oppression, and so on. If our future will be like our past, then we should expect more of the same in the coming decades and centuries. Of course, there will be some differences. The means whereby evil is committed and the precise identity of its victims, for example, may vary. The same is true of atrocities committed in the twentieth century compared to those committed in the eighteenth century, yet the contours of evil remain, regardless of whether the relevant actions are carried out with aid of the gash chamber or the guillotine. If we find a pattern of evil in our past (and present), then it is plausible to expect that pattern to continue into our future, unless there is a good reason to think otherwise. We do find a pattern of evil in our past (and present), and there is no reason I can see to think that will not continue. We should therefore expect more evil in the future.

Clearly this is a pessimistic conclusion. I expect the future to contain evil actions. An optimistic outlook would expect the reverse, or at least it would expect evil actions to wane in their frequency or intensity. More specifically, this is an expectation of catastrophe in both senses specified above. First, we can reasonably expect these evil actions to cause catastrophic harm, given the severity and extent of the harm caused by evil actions. Second, these actions will, by their very nature, involve a catastrophic failure on the part of the moral agents who commit evil actions. This gets us closer to the sort of pessimism I am defending in this book, namely an expectation of catastrophe.

Further, at least some of the evil actions of the future are likely to be ecological in nature. We have already seen this in the case of climate obstructionism, which we can reasonably expect to continue for some time, even if it takes new forms. Another possible evil action is the use of nuclear weapons, of which there is a non-trivial risk (Baum 2015). Among other things, this could be ecologically devastating, both for human and non-human beings, whether it be a large-scale nuclear exchange or something more limited. To take another example, we can reasonably expect wealthy, high-emitting countries in the future to cause ecological devastation for poor countries while refusing to offer any meaningful assistance to them. For instance, sea-level rise will likely turn many people from poor, low-emitting countries into climate refugees (Hauer et al. 2020). If the future resembles the past, then rich, high-emitting countries will do almost nothing to help such refugees, even though rich, high-emitting countries will have been causally responsible for their plight. This would be in keeping with the current and past behavior of rich countries.

In short, humanity's penchant for engaging in evil actions will likely have far-reaching ecological impacts in the future, and this provides yet another reason to be an ecological pessimist.

Is Moralizing Pointless?

As a last objection to all this, one might press the claim that all this talk of evil is socially and politically pointless. This might, but need not, be motivated by a certain strand of Marxism that sees morality as nothing more than ideology (see Buchanan 1987). Just as it is allegedly useless to lecture the capitalist on their moral failings, so it is useless to accuse agents of performing evil actions. Perhaps everything I have said in this chapter is just pointless moralizing.

In truth, I am sympathetic to the objection in general. It just misses the mark in this case. I do not claim that judgments of evil will help us solve practical problems. For instance, I do not assume that agents of evil can be dissuaded from their actions by presenting them with a sound moral argument, nor do I suppose that pressing the charge of evil against (say) politicians and share-holders will motivate them to change their ways. It may well be that ascriptions of evil are pointless in practical terms, at least when it comes to altering behavior. To be clear, I am not certain that such ascriptions are useless, but let us assume that the objector is right about that.

This does not threaten what I have attempted to do in this chapter. I simply think it is true that certain actions qualify as evil and that some evil actions are of an ecological nature. Perhaps pointing this out has zero social-political utility, particularly when it comes to improving the world. That fact would not render false, for example, the claim that climate obstructionism is evil. Of course, I may be mistaken in that claim for other reasons, but the utility of the claim and the truth of the claim are distinct matters. In other words, although I have nothing against trying to change the world, this chapter has sought only to interpret it.

Moreover, the spirit of the objection fits nicely with the pessimistic stance I advocate. Humanity has shown itself throughout history to be comfortable with the performance of evil actions. That is regrettable, but it would be naive to think that philosophy can change that about us. I do not deny that philosophy can have practical significance for certain individuals, informing what they believe and value, but it is not tenable to think that philosophical reasoning can turn humanity in general toward the good, so to speak. In particular, I am pessimistic about the influence moral reasoning can have on the actions of most persons, and thus it is to be expected that arguments regarding evil will, for the most part, have little or no social-political utility. Again, this is just another reason to be pessimistic about our ecological prospects.

Notes

1 Benatar makes a similar point about finding necessary and sufficient conditions for what counts as "meaning" in human life (Benatar 2017, 18).
2 For a useful overview of the concept of evil, see Calder (2022).
3 One thing to notice is that not all forms of obstructionism are morally bad or even morally questionable. Suppose that some political party uses parliamentary tricks to obstruct the agenda of an opposing party. I assume this can be a morally good thing, depending on the circumstances. Our evaluation should depend on whether, say, the thwarted party had sought to enact apartheid. For this reason, I will not argue that obstructionism is evil as such.
4 See also Blosser (1987) and (1995), *Scheler's Critique of Kant's Ethics*.

References

Baum, Seth D. "Confronting the Threat of Nuclear Winter." *Futures* 72 (2015): 69–79.

Benatar, David. 2017. *The Human Predicament*. New York: Oxford University Press.

Blosser, Philip. "Moral and Nonmoral Values: A Problem in Scheler's Ethics." *Philosophy and Phenomenological Research* 48, no. 1 (1987): 139–143. https://doi.org/10.2307/2107712.

Blosser, Philip. *Scheler's Critique of Kant's Ethics*. Athens, OH: Ohio University Press, 1995.

Broome, John. 2012. *Climate Matters: Ethics in a Warming World*. London: W. W. Norton.

Buchanan, Allen E. "Marx, Morality, and History: An Assessment of Recent Analytical Work on Marx." *Ethics* 98, no. 1 (1987): 104–136.

Calder, Todd. "Is Evil Just Very Wrong?" *Philosophical Studies* 163, no. 1 (2013): 177–196.

Calder, Todd. "The Concept of Evil," In *The Stanford Encyclopedia of Philosophy*, edited by Edward N. Zalta and Uri Nodelman, 2022. https://plato.stanford.edu/archives/win2022/entries/concept-evil/.

Card, C. *The Atrocity Paradigm: A Theory of Evil.* Oxford: Oxford University Press, 2002.

Formosa, P. "Evils, Wrongs and Dignity: How to Test a Theory of Evil." *Journal of Value Inquiry* 47, no. 3 (2013): 235–253.

Formosa, P. "Different Substantive Conceptions of Evil Actions." In *Routledge Handbook of the Philosophy of Evil*, edited by Thomas Nys and Stephen de Wijze. London and New York: Routledge, 2019, pp. 256–266.

Hauer, Mathew E., Elizabeth Fussell, Valerie Mueller, Maxine Burkett, Maia Call, Kali Abel, Robert McLeman, and David Wrathall. "Sea-Level Rise and Human Migration." *Nature Reviews Earth & Environment* 1, no. 1 (2020): 28–39.

Hoggett, Paul. "Ressentiment and Grievance." *British Journal of Psychotherapy* 34, no. 3 (2018): 393–407.

IPCC. *Climate Change 2022: Impacts, Adaptation, and Vulnerability.* Contribution of Working Group II to the Sixth Assessment Report of the Intergovernmental Panel on Climate Change [H.-O. Pörtner, D. C. Roberts, M. Tignor, E. S. Poloczanska, K. Mintenbeck, A. Alegría, M. Craig, S. Langsdorf, S. Löschke, V. Möller, A. Okem, B. Rama (eds.)]. New York: Cambridge University Press, 2022.

Kekes, J. "The Reflexivity of Evil." *Social Philosophy and Policy* 15, no. 1 (1998): 216–232.

Kelly, Michael R. "Envy and Ressentiment, a Difference in Kind: A Critique and Renewal of Scheler's Phenomenological Account." In *Early Phenomenology: Metaphysics, Ethics, and the Philosophy of Religion*, edited by Brian Harding and Michael Kelly. London: Bloomsbury Academic, 2016, pp. 49–66.

Matthews, Christopher, and Collin Eaton. "Inside Exxon's Strategy to Downplay Climate Change." *Wall Street Journal*. 14 September 2023. https://www.wsj.com/business/energy oil/exxon-climate-change-documents-e2e9e6af

Meltzer, Bernard N., and Gil Richard Musolf. "Resentment and Ressentiment." *Sociological Inquiry* 72, no. 2 (2002): 240–255.

Morelli, Elizabeth Murray. "Ressentiment and Rationality." *The Paideia Archive: Twentieth World Congress of Philosophy* 16 (1998): 80–86.

Nietzsche, Friedrich. *The Genealogy of Morals in Basic Writings of Nietzsche.* Translated and edited by Walter Kaufmann. New York: Random House, 2000, 437–599.

Nolt, John. "How Harmful Are the Average American's Greenhouse Gas Emissions?" *Ethics, Policy and Environment* 14, no. 1 (2011): 3–10.

Oreskes, Naomi, and Erik M. Conway. 2010. "Defeating the Merchants of Doubt." *Nature* 465, no. 7299 (2010): 686–687.

Oreskes, Naomi, and Erik M. Conway. 2011. *Merchants of Doubt.* New York: Bloomsbury.

Poellner, Peter. *Value in Modernity: The Philosophy of Existential Modernism in Nietzsche, Scheler, Sartre, Musil.* Oxford: Oxford University Press, 2022.

Reginster, Bernard. "Nietzsche on Ressentiment and Valuation." *Philosophy and Phenomenological Research* 57, no. 2 (1997): 281–305.

Russell, Luke. 2007. "Is Evil Action Qualitatively Distinct from Ordinary Wrongdoing?" *Australasian Journal of Philosophy* 85, no. 4 (2007): 659–677.
Russell, Luke. 2014. *Evil: A Philosophical Investigation.* Oxford: Oxford University Press.
Scheler, Max. *Ressentiment.* Translated by Lewis Coser and William Holdheim. Milwaukee, WI: Marquette University Press, 1994.
Scheler, Max. "Ordo Amoris." In *Selected Philosophical Essays.* Translated and edited by David R. Lachterman. Evanston, IL: Northwestern University Press, 1973, 98–135.
TenHouten, Warren D. "From Ressentiment to Resentment as a Tertiary Emotion." *Review of European Studies* 10 (2018): 49.

4 Objections to Ecological Pessimism

Unsurprisingly, my ecological pessimism is subject to certain objections. Here I wish to reply to what I consider the most pressing complaints. The point of this is not merely to defend my view but also to expand and clarify my account.

Objection: Ecological Pessimism Misses the Mark

First, it might be objected that the position I defend is not really *ecological* pessimism at all but something else. One form of this objection is as follows: Although climate change or ocean acidification might be devastating for humanity and many non-human species, the ecology of earth will survive, giving rise to species in the future that will flourish in new ecological niches. In other words, the objection presses the idea that I have been discussing only catastrophes for certain species and perhaps certain ecosystems. A genuine ecological catastrophe would involve the devastation of the earth's ecology as such, not merely a radical change to that ecology. So, the objector might continue, the author of the present book should drop any talk of ecological catastrophe and adopt a more fitting conceptual framing.

Initially, we might accuse someone pushing this objection of semantic trickery, namely stipulating a very narrow definition of "ecological catastrophe." I might respond by saying that my use of the term is perfectly recognizable. Yes, most of the cases I discuss do not involve the collapse of earth's ecology altogether, but they have obvious ecological features (e.g., in their causes and effects). For instance, we are examining changes in the natural environment that have catastrophic impacts not just on human societies but also on various aspects of non-human nature. Using the term "ecological catastrophe" is therefore not idiosyncratic.

Yet there is more to this objection than a trivial semantic complaint. From the point of view of earth's biosphere, things likely will be fine for another billion years or so, in the sense that there will be a functioning

DOI: 10.4324/9781003570523-5

and vibrant biosphere. From this point of view, does it matter precisely which species and ecosystems happen to be around? Looking at the past, many species have gone extinct during earth's history. Some past events were plausibly catastrophic for trilobites but not for life on earth in general. Indeed, it was a necessary condition for the coming into existence of the subsequent species that populate the current biosphere that many of us value. Thus, aside from the semantic question of how to use the term "ecological catastrophe," there is a more interesting question. Every catastrophe must be a catastrophe *for* some subject. As we saw in Chapter 1, a supernova in otherwise-empty space is not catastrophic, but that same event would be catastrophic if it occurred next to an inhabited planet.[1] Granted that some event is catastrophic for existing species and ecosystems, is it perhaps non-catastrophic, or even a good thing, for the biosphere?

Again, there is something to this line of thinking, but it does not pose a problem for my position. For one thing, some of the events I have in mind would be catastrophic for the biosphere. Large-scale nuclear war might devastate life on earth, as might a runaway greenhouse effect causing extreme climate change. But I admit that these are extreme cases, and I could not adequately defend the claim that either one of them is likely.[2] Most of the potential catastrophes I have in mind would leave the biosphere intact and would permit a full recovery, given enough time.

More importantly, this objection allows me to clarify my position. The position I defend expects catastrophe for the species and ecosystems that currently populate our planet. This could take the form of species extinction, the collapse of ecosystems, and harm to individual organisms on a massive scale. Even if life on earth survives and flourishes in the wake of these events, they are still recognizably catastrophic for the victims. If we do value the current constitution of life on this planet, then those catastrophes are to be lamented. We can recognize the badness in them even if they prove to be necessary for the development of new species and ecosystems that we might also value if only we could be around to see them. We can make a similar point looking backward. Present-day mammals do not have much to complain about when it comes to the Cretaceous–Paleogene extinction event, but it obviously was catastrophic for the many species decimated at the time. The fact that life recovered does not change that. Another way of putting this is to say that earth's biosphere did not suffer a catastrophic failure, but various species, ecosystems, and individual organisms did.

In short, we can answer the objection as follows. Why be an ecological pessimist if the biosphere survives and flourishes, albeit in a radically different state? Because that transition would involve massive ills for the constituents of the current biosphere. Perhaps the wholesale destruction of life on earth would be even worse, but the events I have in mind are sufficiently bad to count as catastrophes.

Objection: Ecological Pessimism Overlooks Reasons for Optimism

Another objection is simply that the ecological pessimist ignores many reasons to be optimistic (see Willow 2023). First, from a social-political point of view, there have been environmental successes in the past, such as the Montreal Protocol, which has been largely effective in reducing ozone depletion (Gonzalez et al. 2015). There has been some progress in addressing climate change. We have the UNFCCC, which at least provides a forum and a starting point for possible climate action. The IPCC has been a great success when it comes to collecting and disseminating knowledge regarding climate change. Some countries have taken serious action in reducing greenhouse gas emissions. Even in the United States, some politicians now acknowledge that a potentially society-destroying climate crisis is perhaps worthy of our attention. Leaving climate change to the side, there has been environmental progress in some parts of the world when it comes to reducing air and water pollution, preserving species, and protecting wilderness.

Unfortunately, these successes have been few in number, at least when compared to the many problems we face. When stacked against the ecological ills that we may currently observe or reasonably expect in the future, marginal progress on some issues does not warrant general optimism. Even the examples just mentioned have their dark sides. The Montreal Protocol is remarkable because of the rarity of such success. Progress on climate change, although real, has fallen far short of what is needed. Of course, the ecological pessimist can admit that there has been some progress on some issues. That is compatible with thinking that, on the whole, we are still likely headed for catastrophe.

Let us consider another version of the objection. Although it is true that humanity has made little progress in averting an ecological crisis so far, perhaps the future will be unlike the past. It is possible that new generations will care much more about environmental issues than current and past generations. This could lead to a social-political change that might power meaningful progress. I admit this possibility, but it appears unlikely. For one thing, we are already committed to certain risks, such as climate change due to past emissions. Even if future generations undertake substantial risk mitigation, there are limits to what can be done. Such efforts may come too late. On the other hand, the required policies may require drastic changes to the status quo, arguably amounting to a global revolution. For instance, imagine a popular movement seeking to reorder the global economy such that it served the needs of human beings in general rather than only the ultra-wealthy. If it became influential, that movement would meet brutal opposition, political at first and violent if necessary. This does not bode well for those who wish to avert ecological catastrophe.

Another target of potential optimism is technology, as one sees among so-called "ecomodernists" (Karlsson 2020; Symons 2019). Perhaps some technological achievement in the future will greatly reduce the probability of ecological catastrophe. This may take the form of a general "techno-optimism." As a provisional definition, Danaher suggests that "techno-optimism is the stance that holds that technology plays a key role in ensuring that the good prevails over the bad" (Danaher 2022, 8). Perhaps this does not fully capture the negative connotation often associated with the term. We might think that a genuine techno-optimist is someone who expects technology to save us in some way, to provide a kind of salvation. Admittedly, this raises the worry that the charge of techno-optimism will be deployed as a strawman. Outside of some odd people in Silicon Valley, does anyone really believe in technological salvation? Regardless of the precise definition, however, it is reasonable enough to say that a proponent of techno-optimism is likely to claim that future technologies will provide solutions to some of the ecological problems we face. Even this relatively modest prediction is uncertain and open to critique (Alexander and Rutherford 2019; Keary 2016). Likewise, even if one accepts merely that "technology plays a key role in ensuring that the good prevails over the bad," this does not automatically ensure that solutions will be found for the specific issues that might bring about ecological catastrophe.

There is plenty of room for skepticism about specific technologies. As many have pointed out, the rosiest emissions projections from the IPCC rely on the large-scale use of "negative emissions" technologies, which are unproven and carry their own costs (Lenzi 2021). It is not known whether such technologies could be deployed at the scale needed to counter-balance anthropogenic emissions. That might turn out to be impractical due to land-use pressures or costs (Fuss et al. 2018). Yet even if the wide-scale use of negative emissions technology is technically and economically feasible, decision-makers might fail to pursue their deployment. A glance at the United States Congress should be sufficient to ground the worry that decision-makers will not act rationally. To be clear, I am not objecting to the IPCC studying the possible use of negative emissions technologies, but it would be foolish to rely on unproven technologies as a matter of policy. Doing so would count as an indefensible type of techno-optimism.

The most plausible appeals to technological intervention are those that think some technology might prove useful as one tool among many and only under specific conditions. For example, in previous work, I have been friendlier to geoengineering than most ethicists, but that is hardly praise (see Pamplany et al. 2020). Roughly put, my view is that, in some future scenario, geoengineering in the form of stratospheric aerosol injections might be part of the least bad climate policy. However, if we reach that point, it is because we have fundamentally failed to address climate change in an ethically appropriate way, such as by reducing our emissions. In an ideally just world, we would have initiated substantial cuts to our emissions decades ago, and in that case, geoengineering would have no place. Of course, the real world took a different course. Perhaps, then, geoengineering could be used at some point to ease the harms of climate change, even while introducing problems of its own (Morrow and Svoboda 2016). If geoengineering is ethically permissible in the future, it will be because past ethical wrongdoing (e.g., refusing to mitigate emissions) produced a bad situation in which drastic actions might be part of our least bad option (Svoboda 2017). Moreover, I think the use of geoengineering, even if ethically permissible in some cases, should be deeply regretted (Svoboda 2015). This position is not a case of techno-optimism. Even though it allows that technological intervention can become worthy of consideration in bad times, I do not think it will provide anything resembling salvation.

In general, I have never encountered a living researcher who thinks geoengineering would secure paradise on earth. I say "living researcher" because historically some scientists were excited about the prospect of geoengineering, although this evaluation was based on naively optimistic speculation (see Fleming 2010). At best, current proponents of researching geoengineering think it might be a useful yet "imperfect" tool in responding to climate change (Keith et al. 2010). It is arguably inapt to call this techno-optimism, at least not in a pejorative sense. The idea that technology might play some helpful role in the climate crisis is a rather modest position, especially when that is coupled with the recognition that emissions mitigation is to be preferred to technological intervention.

Objection: The Future Is Deeply Uncertain

It might be argued that the ecological future is "deeply" uncertain and that we therefore lack any legitimate basis for making predictions about that future. This would undermine the pessimistic claim that ecological catastrophe is likely to occur in the future. The same, presumably, would hold for an optimistic prediction. The reason is that any such catastrophe depends on complex systems that are, at least in some cases, poorly understood. Specifically, different studies of the same risk provide inconsistent probability distributions. A given study may find the risk of some catastrophe is more or less probable than some other study finds, sometimes by a wide margin. We are therefore not in a position to say whether or not the catastrophe is likely to occur in the future. We simply do not know, and claims about what is likely to occur in such contexts will be unwarranted. Generalizing from this, we might think that ecological pessimism is unwarranted, because it fails to deal with the deep uncertainty of various phenomena, and yet it purports to tell us what is likely to occur.

This second-order uncertainty is referred to as "deep" or "Knightian" uncertainty (see, for example, Haas et al. 2023). Unlike first-order uncertainty, where we can assign probabilities to various outcomes, under deep uncertainty, we are uncertain about those probabilities themselves. In other words, in some cases, we might not know whether an outcome is likely or unlikely. Consider a shutdown of the Atlantic Meridional Overturning Circulation (AMOC), something that would plausibly count as an ecological catastrophe. Under climate change, there is some risk of this occurring, but competing studies do not agree on how likely this is under possible future conditions (Chen et al. 2019; Reintges et al. 2017). The probability of the AMOC shutting down in the future is itself a matter of uncertainty. Note that this holds even when various conditions, such as global average surface temperature and precipitation, are assumed. Some of these assumed values are deeply uncertain themselves. For instance, how much future warming we can expect given a doubling of carbon dioxide in the atmosphere relative to pre-industrial levels—which is often referred to as climate sensitivity—is unclear, with different modeling studies providing divergent probability density functions for various degrees of warming. Of course, the probability of the AMOC shutting down in the actual future depends in part on what climate sensitivity turns out to be, which is itself deeply uncertain. This creates problems for those who wish to make predictions about the future, especially for those relying on integrated assessment models, which rely on both geophysical and economic projections (Frisch 2013).

There are several plausible responses to this objection. First, while the probability of some outcome might be deeply uncertain, it nonetheless could be clear that the outcome is likely. This is because all available studies might agree that the probability of the outcome is greater than .5, even while they disagree about its exact value. In such a case, we would be warranted in saying that the outcome is likely, although we might not be warranted in saying much more than that. For this reason alone, the concerns just expressed might not hold in all cases. Whether deep uncertainty is a problem for the ecological pessimist will depend on what the various probability density functions of the relevant studies actually say. Admittedly, this would be a convenient state of affairs for the ecological pessimist, but it is a realistically possible one.

Second, and more importantly, the central claim of ecological pessimism does not depend on the probability of any specific phenomenon. To say that ecological catastrophe is likely to occur is to say that *some* ecological catastrophe is likely. We need not specify exactly which one, nor must we be committed to the view that any particular catastrophe is more likely than not. Suppose that, due to deep uncertainty, it is not clear that any particular catastrophe is more likely to occur than not. The pessimists can accept that, for they claim only that it is more likely than not for at least one of these phenomena to occur. For this reason, deep uncertainty poses a serious problem for me only if it turns out that, for *every* potential ecological catastrophe, there is good evidence (e.g., from a modeling study) to assign a very low probability to it. I say "very low" because that is needed in order for the overall probability of *just one* of those catastrophes occurring to remain below .5. Given the many forms that ecological catastrophe might take and the ample time available, it is very unlikely that this overall probability is below .5. We can therefore make the pessimistic claim despite the presence of deep uncertainty regarding many phenomena.

Objection: Ecological Pessimism Is Alarmist

Environmentalists are sometimes charged with being alarmist. The charge is ambiguous. Clearly, there are cases in which raising an alarm is warranted, such as urging people to leave a burning building, so problematic alarmism must involve something more than merely sounding an alarm. Usually, the alleged alarmist is thought to exaggerate risk and thereby create untoward personal or social effects, such as poor economic decisions or needless anxiety. If someone claims that the world's oceans are on track to boil away in the next decade, we can rightly charge them with alarmism, because this claim greatly exaggerates a risk and, if taken seriously, could lead to unfortunate outcomes: panic, anxiety, and foolish expenditures, for example. By contrast, consider someone who claims that there is a small probability of the world's oceans boiling away in the very distant future (Goldblatt and Watson 2012). This claim is not plausibly taken to be alarmist, because although it acknowledges the possibility of a very serious phenomenon, it does not exaggerate that risk, nor is it likely to create the personal and social problems of predicting imminent disaster.

The question, then, is whether the ecologically pessimistic view both exaggerates risks and creates untoward effects. I think it does not. Although holding that catastrophe in some form is likely to occur sounds alarming in some sense, there is good evidence to support it, as I have argued in previous chapters. Contrast that with the doom-sayer who predicts the near-term disappearance of the oceans. If that is even possible, it is extremely unlikely to occur, and there is clearly insufficient evidence to support such a prediction. My view is much more modest, as it merely expects some catastrophe to occur in the future. I do not claim to know which specific ones will occur, nor do I claim to know when they might occur. Once again, there is vagueness built into my position, particularly when it comes to the specifics and timing of possible catastrophes. This vagueness may be frustrating, but it is unavoidable. No one is in an epistemic position to know when, where, and how specific events in the future might occur. At best, we have evidence that supports probabilistic judgments. If the world warms by (say) five degrees, it is reasonable to expect the West Antarctic Ice Sheet to collapse at some point, but no one can know precisely when. That expectation is not alarmist, because it does not involve any exaggeration of risk. It may produce unwelcome personal and social effects, a matter I address in response to other objections, but recognizing an unwelcome truth is not alarmism.

Let me make one more point. Although perhaps not a necessary component of alarmism, genuinely alarmist positions often exaggerate the risk of imminent phenomena. Our doom-sayer above claims that the oceans will disappear in the next decade. If they had instead predicted that this would occur in a million years, perhaps we would not consider them to be an alarmist. This may be because of certain psychological and social facts about us. A near-term catastrophe is much more worrying than an equally likely long-term catastrophe, and that shortness of time may lead us to make bad decisions. There is nothing in my position that appeals to imminent catastrophe. Although it is possible that catastrophe will occur in the next decade, it is more likely to occur in the more distant future, perhaps after all of us who might read this are dead. This is not the sort of claim that we can plausibly accuse of alarmism. Climate scientists have long warned of dangers that, at least from the human point of view, dwell in the distant future. Their temporal distance perhaps explains why many of us remain unalarmed.

Objection: Ecological Pessimism Undermines Individual Action

Another objection is practical in nature, pointing to ecological pessimism's alleged tendency to undermine action. This objection can take many forms, but the basic idea seems plausible enough: If we accept the pessimistic attitude, then we are less likely to undertake certain worthwhile actions. I will review different forms of the objection below. What they all have in common is that they reject ecological pessimism not because it is false but rather because it supposedly has unwelcome practical effects.

We may begin with personal versions of the objection. First, one might think that accepting the pessimistic attitude would damage an individual's *motivation* to perform worthwhile actions, such as those that are thought to promote sustainability or other environmentally desirable ends. Imagine an individual who was a model ecological citizen before learning about, and becoming convinced by ecological pessimism. As a consequence, she has lost some or all motivation to reduce her greenhouse gas emissions, recycle, conserve energy, and so on. Of course, whether adopting the pessimistic attitude would have this impact on motivation, and to what degree, is an empirical matter. No doubt it would vary somewhat among different individuals.

Instead of motivation, an objector might focus on *reasons*. On this variety, one might say that accepting ecological pessimism weakens or removes some of the reasons one has to perform worthwhile actions. If ecological catastrophe is likely to occur regardless of what I do, then why should I care about (say) sustainability in my personal life? I might decide that all things considered, I have more reason to act in ways that serve my own comfort and convenience. If not for the pessimistic attitude, I would (perhaps) recognize reasons to prioritize goals of sustainability or the like over my personal interests. The difference here is that no claim is made about my actual motivation. This reasons-based version of the objection therefore has more normative force than the motivation-based version. Whether or not my motivation is reduced as an empirical matter of fact, the objection claims that accepting ecological pessimism will weaken the justification for my acting in desirable ways. Why should that be a problem? Perhaps because there are reasons for an individual to live sustainably, but the pessimistic attitude fails to recognize that.

There are several ways of responding to this objection. First, my position is that ecological pessimism is true, because ecological catastrophe is likely to occur in the future. The psychological or normative ramifications of accepting that position are simply irrelevant to its truth or falsity. Perhaps accepting the truth in this case has unwelcome results. That would just make it a true belief with unfortunate consequences. Surely it would be nice if truth always converged with other things we value, but that is not the case. John Keats says, "Beauty is truth, truth beauty,—that is all / Ye know on earth, and all ye need to know" (Keats 2001, 240). Nietzsche's stance is more plausible: "For a philosopher to say, 'the good and the beautiful are one,' is infamy; if he goes on to add, 'also the true,' one ought to thrash him. Truth is ugly" (Nietzsche 1968, 435). Whatever one thinks about Nietzsche's view, it is clear that many truths in our world are ugly and terrible. It is true that thousands of children suffer and die every day from easily preventable causes, that injustice often goes unpunished and acknowledged, and that ecological catastrophe is likely to occur.

However, to borrow again from Nietzsche, we might ask why truth should take priority over other values: "Suppose we want truth: *why not rather* untruth? and uncertainty? even ignorance?" (Nietzsche 2000: 199). Initially, this might seem to be a foolish question, but it is actually very important. "Untruth" includes not just falsehood but also phenomena that are neither true nor false, such as desires, preferences, and emotions. A life devoted solely to the pursuit of truth would miss the valuable aspects of these phenomena, such as the emotional connections pertaining to interpersonal relationships. Usually, we can value this sort of "untruth" while also maintaining respect for truth. This is because attitudes that are not truth-apt do not conflict with attitudes that are truth-apt. For instance, I can feel an emotional attachment to a friend while also recognizing the fact that they have various flaws. Matters are more interesting when we turn to cases of outright falsehood and ignorance.

In fact, many people do prefer falsehood to truth in various contexts of human life: sports, politics, self-image, and interpersonal relations, for example. Fans will deny obvious facts that are detrimental to their team and gladly accept mistaken officiating calls that favor their team. Voters support candidates who tell them what they want to hear, no matter how outlandish. People have wildly inaccurate beliefs, skewing both positive and negative, about their own physical and intellectual qualities. Some believe their children can do no wrong while others greatly exaggerate the vices of their children. Here we have cases of untruth that directly conflict with truth. We cannot have both in this case, so what are we to prefer? Philosophers may take it as obvious that we ought always to prefer truth, but why? Nietzsche thinks that preference is just a "prejudice" of philosophers. All of this opens the way to wondering whether ecological pessimism ought to be ignored even if it happens to be true. We might say, "Yes, it is true that ecological catastrophe is likely to occur. But believing that truth is bad, because it undermines certain motivations and/or reasons for worthwhile actions. So we should reject ecological pessimism even though it is true."

For this objection to work, it needs to be the case that the value of those motivations and/or reasons outweighs the value of believing the truth, in some sense of outweighing that remains to be specified. This is doubtful when we consider that false beliefs often carry harmful consequences. Epistemically irresponsible voters can do great harm, including to themselves, by supporting unfit leaders. This can enable war crimes, domestic injustice, and incompetent management resulting in unnecessary death and suffering. An obvious example is the handling of COVID-19 in the United States, where the issue became politicized in some particularly stupid ways. Many functional adults believed obvious falsehoods, spread conspiracy theories, and refused to accept well-supported guidance from public health professionals. This allowed them the pleasure of believing what they wanted and indulging in certain fantasies. The result was a great deal of avoidable death, suffering, and economic damage (Stoto et al. 2022). This is a case in which the benefits of believing the truth obviously outweigh the benefits of believing a falsehood. If it is true that ecological catastrophe is likely, then although rejecting that truth might have some benefits (e.g., a pleasing illusion that our individual actions matter), it also could have harmful consequences—insufficient preparation, for example. Why truth rather than untruth? Sometimes because believing the truth serves our self-interest in avoiding harm.

These considerations suggest another response to the objection, namely that we have ethical reasons, perhaps even an obligation, to believe the truth. Those who held false beliefs about COVID harmed themselves in such cases, but they also harmed others. Although it is foolish to cause oneself needless harm, we *might* think it ethically permissible.[3] Regardless, few will deny that causing harm to others is ethically problematic, especially when that harm results from one's own preference for untruth over truth. My claim is not that merely holding a false belief is ethically impermissible. That would be too strong, as it would include many warranted beliefs that turn out to be false through no fault of the believer. What I have in mind is epistemically irresponsible behavior, such as holding a false belief for ridiculous reasons and thereby causing harm to innocent others. We have an ethical responsibility to seek truth and avoid falsehood, at least in part because doing so, or failing to do so, is tied to whether or not we cause harm to others, say by spreading a deadly disease.

This is akin to William Clifford's "ethics of belief." Clifford tells the story of a man who fails to provide proper maintenance for a ship he owns. As a result, the ship fails at sea and the sailors aboard die. The ship owner did not seek to harm the sailors, nor did he bear them any ill will. Moreover, he genuinely believed that the ship was in serviceable condition. That belief was false, of course, but the important point is that the owner ought to have known better, and he would have known better had he taken the proper steps to inspect the ship instead of assuming it was in fine condition. In short, the man's belief was unwarranted. He believed what he wanted to be true, because that was cheaper and more convenient. Clifford argues, plausibly to my mind, that the fault here is not merely intellectual but also moral (Clifford 2010). We have an ethical responsibility to seek truth, and this at least requires us to take reasonable steps of ensuring that our beliefs are warranted. All of this suggests yet another reason as to why we should prefer truth to untruth.

Finally, we might simply value what Robert Nozick calls "contact with reality" (Nozick 1974, 45). In his classic thought experiment, Nozick asks whether it would be in someone's self-interest to plug into an "experience machine" that stimulates the brain to experience anything that one might wish. If the quality of our experiences is all that matters, then it seems that we should indeed plug into the machine, as that would offer a more reliably good set of experiences than ordinary life, with all its risks and disappointments. Of course, virtually no one thinks it is a good idea to trade ordinary life for the artificial life (but genuine experiences) afforded by the machine. What explains this? Nozick suggests that we place a great deal of value on contact with the real world, something we would lose if plugged into the machine, as our experiences would not refer to anything real. We might say that accepting ecological pessimism is somewhat similar. If ecological pessimism is true, then it accurately describes the (likely) real world, and perhaps we should believe it despite its unwelcome practical effects. Rejecting this true position would be a way of denying what is real.

Objection: Ecological Pessimism Undermines Collective Action

A more concerning issue is that ecological pessimism might interfere with worthwhile collective efforts, such as attempts to reduce risks of catastrophe or to pursue other social goods. Whereas the inaction of a single individual is unlikely to make a difference, collective inaction could have far-reaching consequences. Much of what I say in reply to the previous objection applies here as well. Ecological pessimism simply might be true, and we have reasons to accept the truth even if it has untoward social effects. Let us put that aside, however. As with individual action, adopting pessimism could reduce the collective *motivation* of groups. Again, whether this would be so is an empirical matter. At present it is not clear whether, nor to what extent, an ecologically pessimistic attitude would decrease our collective motivation to pursue worthwhile ends. Perhaps empirical investigation would show that adopting pessimism severely undermines collective motivation, which I admit would be an unwelcome effect, or perhaps it would show that there is little connection. I cannot speculate on that. Instead, I can speak to the collective *reasons* we have for undertaking collective action to reduce ecological risks.

Believing that ecological catastrophe is likely provides no reason for apathy, passivity, or quietism. It would be a mistake to reason that, because catastrophe is likely, there is no point to the collective pursuit of environmental goals. One reason for this is that some catastrophes are worse than others. Even if we cannot avert catastrophe altogether, we can still have moral and prudential reasons for attempting to avert *some* catastrophe. Perhaps, for example, limiting global warming to 2°C is a lost cause, but that does not entail that we should throw up our hands and make no attempt to prevent five degrees of warming. The pessimism I am defending might entail that we are doomed to live in a bad world, but not the worst of all possible worlds. So long as we can prevent the worst from happening, we will have reasons for doing so.

Another reason for the pessimist to eschew apathy, passivity, and quietism is that risks can be reduced. The ecological pessimist is committed to the view that catastrophe is likely, but that judgment is a matter of degree. If some course of collective action can reduce the probability of some catastrophe from .9 to .6, that is a good reason to pursue that course of action, even though it would remain the case that the catastrophe is likely to occur. The pessimistic attitude is perfectly compatible with such efforts.

Again, as a matter of social-psychological fact, I do not know how widespread acceptance of ecological pessimism would impact collective motivation. Perhaps many newly minted pessimists would judge, falsely, that there is no point in attempting to avoid the worst possible world or to reduce risks. That would be unfortunate. It would also be a misapplication of the position I am defending. As we have seen, the ecological pessimist has plenty of reasons to promote and take part in collective actions aimed at environmental and social goods, even if those goods amount to an amelioration of certain ills, an issue to which I return in later chapters.

Objection: Ecological Pessimism Is Really Just Ecological Realism

Another line of attack is to claim that my position is not really pessimistic at all. Although "ecological pessimism" sounds more exciting, perhaps "ecological realism" would be a more appropriate name. Unlike many of the German pessimists (Beiser 2016), my view does not hold that this is the worst of all possible worlds, nor does it even treat bad outcomes as necessary. Rather, my so-called pessimism merely expects catastrophe in a probabilistic sense. One might see this as falling far short of genuine pessimism. As Janaway says about Schopenhauer, human existence "must contain suffering, and cannot be preferable to non-existence. It would even have been better for reality not to have existed. These claims make Schopenhauer a pessimist in a philosophically interesting sense" (Janaway 1994). By contrast, my view is at least compatible with the beliefs that this world need not contain suffering and that existence is preferable to non-existence. On its own then, my account is far less dark than Schopenhauer's. Does it really warrant being called pessimistic? Moreover, most of my arguments for why ecological catastrophe is likely depend upon seemingly contingent facts about human beings and our institutions, as well as the political and economic conditions under which we currently operate. For this reason, perhaps my position is just a type of political realism, urging us to think about our ecological prospects in terms of what is likely to happen rather than in terms of what, ideally, ought to happen.

First, let me note that the issue of realism is distinct from the issue of pessimism (or optimism). The latter involves a value judgment, providing an evaluative assessment of what we can expect. The former does not inherently or necessarily involve a value judgment but instead seeks to identify what is feasible or realistic. Of course, we may assess some feasible option as good or bad, but that judgment is not the same as determining what is feasible in the first place. For instance, in the case of an ongoing war, a realist might correctly note that a lasting peace is not feasible. This would just be a reflection of the social, political, or logistical facts that happen to hold. Now one might be pessimistic about this, seeing the ongoing violence as lamentable, while someone else might be optimistic, seeing (say) political opportunity for himself due to continued fighting. Either way, the realist's judgment is logically independent of any optimistic or pessimistic take on the situation.

With this in mind, I hold that my position is both realistic and pessimistic, and these two features are not in tension. My realism includes thinking that serious, large-scale efforts to reduce ecological risks are socially, politically, and economically infeasible. My pessimism includes thinking that this will likely result in catastrophe. Although related in various, obvious ways, the two views are logically distinguishable. One does not entail the other, for example. We might have lived under luckier conditions, where minimal efforts to reduce ecological risk would be sufficient to avoid catastrophe. Or we might have lived under unluckier conditions, where even the greatest efforts to reduce risk would prove insufficient. My point is just that pessimism is not in competition with realism. Thus, even if it is true that my position is a realist one in some ways, that does not prevent it from also being pessimistic.[4]

Second, although it is true that my pessimism is less severe than that of Schopenhauer and other philosophical pessimists, that merely shows that there are various degrees of pessimism. I think that ecological catastrophe is likely but not certain, and that expectation is supported by contingent facts that could change. Moreover, my pessimism is limited to ecological matters and is therefore compatible with optimism in other areas. Nonetheless, my position expects catastrophe in the future, an expectation that is motivated in part by humanity's widespread moral corruption. It is reasonable to call this "pessimism," even though it is not as encompassing or hopeless as more extreme varieties, for it offers a bleak view of humanity and our ecological prospects.

To conclude this thought, in at least one way my view is more pessimistic than that of Schopenhauer. For him, the nature of will as thing-in-itself makes suffering a necessity. It is unavoidable that the future, like the past, will be bad. In my view, there is no necessity for ecological catastrophe. If we wished, we could do a great deal to reduce the risks of catastrophe. But we don't wish that. Instead, we do very little to avert, untroubled that the world might burn as a result. As I have argued elsewhere (Svoboda 2022), this warrants a rather harsh moral indictment of our species. Agents in Schopenhauer's world would have some excuse. No matter what they do, the will that objectifies itself in them and everything else ensures suffering. If I am right, then human beings have no such excuse. This makes us morally responsible for ecological catastrophe in a way that we could not be in Schopenhauer's world. In short, ecological catastrophe is contingent upon our own action and inaction, not upon some unchanging metaphysical principle. Humanity could be morally decent, but it has chosen a different path.

Notes

1 One might claim that the supernova event is catastrophic for the self-destructing star. Perhaps so, but this does not challenge the relevant point that a catastrophe must be catastrophic for something, in this case, the star itself.
2 Recall that my claim is that *some* ecological catastrophe is likely to occur in the future. I do not claim to know which ones in particular are most likely to occur.
3 On the other hand, and somewhat controversially, we might think self-harm is impermissible, say because we have a duty to respect ourselves.
4 Here is another possible response, although one that I do not emphasize because, in some ways, it runs counter to what I have just said. In general, people can be optimistic or pessimistic about what is realistic or feasible. The philosophical pessimists might claim that they are genuine realists, because they correctly identify what is feasible—e.g., by claiming that suffering cannot be avoided. Optimists, conversely, might hold that they are the true realists—e.g., by claiming that it is feasible to avoid suffering. Here the dispute concerns whether it is the optimists or the pessimists who are mistaken about the nature of what is realistic or feasible. On this approach, which I have not taken above, I could claim that I am a "pessimistic realist," that is someone who thinks our feasible prospects are bleak. This would be another way of arguing for the compatibility of realism and pessimism, thereby sidestepping the objection.

References

Alexander, Samuel, and Jonathan Rutherford. "A Critique of Techno-Optimism." In *Routledge Handbook of Global Sustainability Governance*, edited by A. Hayden, D. Fuchs and A. Kalfagianni. Abingdon and New York: Routledge, 2019, 152–167.

Beiser, Frederick. *Weltschmerz*. New York: Oxford University Press, 2016.

Chen, Changlin, Wei Liu, and Guihua Wang. "Understanding the Uncertainty in the 21st Century Dynamic Sea Level Projections: The Role of the AMOC." *Geophysical Research Letters* 46, no. 1 (2019): 210–217.

Clifford, William Kingdon. *The Ethics of Belief and Other Essays*. Amherst, NY: Prometheus Books, 2010.

Danaher, John. "Techno-Optimism: An Analysis, an Evaluation and a Modest Defence." *Philosophy & Technology* 35, no. 2 (2022): 1–29.

Fleming, James Rodger. *Fixing the Sky: The Checkered History of Weather and Climate Control*. New York: Columbia University Press, 2010.

Frisch, Mathias. "Modeling Climate Policies: A Critical Look at Integrated Assessment Models." *Philosophy & Technology* 26 (2013): 117–137.

Fuss, Sabine, William F. Lamb, Max W. Callaghan, Jérôme Hilaire, Felix Creutzig, Thorben Amann, Tim Beringer, et al. "Negative Emissions—Part 2: Costs, Potentials and Side Effects." *Environmental Research Letters* 13, no. 6 (2018): 063002.

Goldblatt, Colin, and Andrew J. Watson. "The Runaway Greenhouse: Implications for Future Climate Change, Geoengineering and Planetary Atmospheres." *Philosophical Transactions of the Royal Society A: Mathematical, Physical and Engineering Sciences* 370, no. 1974 (2012): 4197–4216.

Gonzalez, Marco, Kristen N. Taddonio, and Nancy J. Sherman. "The Montreal Protocol: How Today's Successes Offer a Pathway to the Future." *Journal of Environmental Studies and Sciences* 5 (2015): 122–129.

Haas, Christian, Henriette Jahns, Karol Kempa, and Ulf Moslener. "Deep Uncertainty and the Transition to a Low-Carbon Economy." *Energy Research & Social Science* 100 (2023): 103060.

Janaway, Christopher. *Schopenhauer*. New York: Oxford University Press, 1994.

Karlsson, Rasmus. "Conflicting Temporalities and the Ecomodernist Vision of Rewilding." In *Non-Human Nature in World Politics: Theory and Practice*, edited by Joana Castro Pereira and André Saramago. Cham, Switzerland: Springer Nature, 2020, 91–109.

Keary, Michael. "The New Prometheans: Technological Optimism in Climate Change Mitigation Modelling." *Environmental Values* 25, no. 1 (2016): 7–28.

Keats, John. *Complete Poems and Selected Letters of John Keats*. New York: Random House, 2001.

Keith, David W., Edward Parson, and M. Granger Morgan. "Research on Global Sun Block Needed Now." *Nature* 463, no. 7280 (2010): 426–427.

Lenzi, Dominic. "On the Permissibility (or Otherwise) of Negative Emissions." *Ethics, Policy & Environment* 24, no. 2 (2021): 123–136.

Morrow, David, and Toby Svoboda. "Geoengineering and Non-Ideal Theory." *Public Affairs Quarterly* 30, no. 1 (2016): 83–102.

Nietzsche, Friedrich. *Beyond Good and Evil*. In *Basic Writings of Nietzsche*. Translated and edited by Walter Kaufmann. New York: Modern Library, 2000.

Nietzsche, Friedrich. *The Will to Power*. Translated by Walter Kaufmann and R. J. Hollingdale. Edited by Walter Kaufmann. New York: Vintage, 1968.

Nozick, Robert. *Anarchy, State, and Utopia*. United Kingdom: Basic Books, 1974.

Pamplany, Augustine, Bert Gordijn, and Patrick Brereton. "The Ethics of Geoengineering: A Literature Review." *Science and Engineering Ethics* 26 (2020): 3069–3119.

Reintges, Annika, Thomas Martin, Mojib Latif, and Noel S. Keenlyside. "Uncertainty in Twenty-First Century Projections of the Atlantic Meridional Overturning Circulation in CMIP3 and CMIP5 Models." *Climate Dynamics* 49 (2017): 1495–1511.

Stoto, Michael A., Samantha Schlageter, and John D. Kraemer. "COVID-19 Mortality in the United States: It's Been Two Americas from the Start." *PLoS One* 17, no. 4 (2022): Article e0265053.

Svoboda, Toby. "Geoengineering, Agent-Regret, and the Lesser of Two Evils Argument." *Environmental Ethics* 37, no. 2 (2015): 207–220.

Svoboda, Toby. *The Ethics of Climate Engineering: Solar Radiation Management and Non-Ideal Justice*. New York: Routledge, 2017.

Svoboda, Toby. *A Philosophical Defense of Misanthropy*. New York: Routledge, 2022.

Symons, Jonathan. *Ecomodernism: Technology, Politics and the Climate Crisis*. Oxford: John Wiley & Sons, 2019.

Willow, Anna J., ed. *Anthropological Optimism: Engaging the Power of What Could Go Right*. New York: Routledge, 2023.

5 Environmental Philosophy as a Way of Life

Suppose one accepts my arguments in the preceding chapters and comes to adopt an ecologically pessimistic outlook. How should that person live? This question is, of course, practical in nature. Asking it presumes that adopting the outlook in question is not merely a theoretical matter. Perhaps it is possible for someone to judge that we are likely driving ourselves into ecological catastrophe without that judgment influencing her desires, choices, or actions to the slightest degree. But we can ask whether that would be a rational or desirable outcome. Perhaps the answer is yes, but we need to consider the issue more carefully. In the following chapter, I will consider specifically how one might live as an ecological pessimist. However, I will do that within a framework that I lay out in the current chapter, namely environmental philosophy as a way of life.

Environmental philosophy is particularly well-suited to facilitate a revival of a philosophical art of living, or the practice of philosophy as a way of life. The notion that philosophy involves the practice of living well is most often associated with Hellenistic figures (e.g., Epicurus, Epictetus, and Seneca), but it is also present in some modern philosophical writers (e.g., Thoreau). However, despite interest in this tradition of philosophy from the likes of Michel Foucault, Martha Nussbaum, and Pierre Hadot, the practice of philosophy as a way of life is virtually absent at the present time.

In this chapter, I argue both that philosophy as a way of life is a tradition worth reviving and that environmental philosophy is a promising branch of philosophy to enact this revival. First, I sketch what constitutes philosophy as a way of life, which includes both some conception of the good life and an array of spiritual exercises that assists one in living according to that conception. I then discuss a connection between possessing virtue and leading the good life, a connection of great importance to ancient and modern practitioners of philosophy as a way of life. Next, I offer an argument for why this tradition of philosophy is worth reviving at the present

DOI: 10.4324/9781003570523-6

time. The remainder of the chapter is devoted to exploring the prospects for a distinctively environmental approach to philosophy as a way of life.

Given its emphasis on environmental virtue and its rich resources for developing spiritual exercises, I argue that environmental philosophy as a way of life is both a robust and attractive option compared to what we might call "purely theoretical" approaches to environmental philosophy. My argument is that, because environmental philosophy involves normative claims for human beings, we need some way of internalizing the relevant norms and acting in accordance with them. This is precisely what environmental spiritual exercises provide. Purely theoretical approaches tend to fall short in this respect, for they specify norms without providing means to internalize, enact, or care about them. This is an aspect of what Lisa Kretz calls the "theory-action gap." Using the example of climate change, she notes that the ethical case for decisive action is overwhelming, and yet many of us fail to act accordingly (Kretz 2012). Others have made similar claims about what I am calling purely theoretical approaches, including that these overlook the psychology of motivation (Booth 2009) and pay insufficient attention to the philosophy of action (Coeckelbergh 2015; Goralnik and Nelson 2011). The general problem for purely theoretical approaches is that they do not address the theory–action gap. Such approaches might provide unassailable arguments regarding how we ought to act and yet have very little impact on how human beings end up acting, even in the case of those who understand and accept the relevant arguments. I shall argue that environmental philosophy as a way of life offers an attractive way to bridge this gap, and for two reasons: first, it provides techniques (i.e., spiritual exercises) for internalizing relevant norms, such as those advocated by some theory; second, it takes seriously the idea that certain attitudes and actions regarding the environment contribute to one's own flourishing, thus providing some motivation to take on those attitudes and actions.

Philosophy as a Way of Life

In order to live well as a human being, it seems helpful to have both (1) some conception of the good life and (2) some array of practices or "spiritual exercises" (Hadot 1995) whereby one is able (or becomes able) to lead such a life. Historically, philosophers have provided resources to assist in both (1) and (2), presenting competing views of the good life and developing practices whereby one actually could lead or at least approximate some specified good life. Yet while contemporary philosophers have continued to think about what constitutes the good life, very few carry on the ancient tradition of thinking about spiritual exercises, and fewer still advocate practicing them (although cf. Irvine 2008). Following Hadot, by "spiritual exercises," I shall mean endeavors "intended to effect a modification and a transformation in the subject who practiced them" (Hadot 2002, 6). We might also call such practices "ascetic exercises," understood in the Greek sense of *askesis* as transformation, for the point of performing them is to bring about a transformation in oneself, one to be conducted in accordance with some conception of the good life. Spiritual exercises may include practices such as meditation, physical exercise, diet regimens, manual labor, journal-writing, self-examination, and so on. What makes these *spiritual* exercises is that their goal is to bring about an internal change in oneself, rather than an external or material change about oneself. As Arnold Davidson says, these exercises "were *spiritual* because they involved the entire spirit, one's whole way of being" (Davidson 1995). The term "spiritual" does not require that these exercises be religious in nature (although they can be) but only that they be concerned with an inner transformation of the subject practicing them. This is why the term "ascetic" or "transformative" exercises would serve just as well, but I use the more common "spiritual exercises," as employed by Hadot and others.

Philosophers are not the only ones who have pursued (1) and (2), of course. Most obviously, various religious and monastic traditions contain both rich conceptions of the good life and sophisticated spiritual exercises. Given that these endeavors are not exclusive to philosophy, something further must be said regarding what distinguishes philosophy as a way of life from non-philosophical (or not purely philosophical) approaches to (1) and (2). As will become clear from the cases discussed below, a distinctively philosophical approach to (1) and (2) involves rational reflection on both what the good life is and how one might cultivate oneself to lead the good life. I will not attempt to define what exactly counts as rational reflection. There are competing conceptions of rationality available, and this controversial issue cannot be resolved with adequacy here. However, we can expect that a rationally reflective approach to (1) and (2) will include argumentation, appeals to coherence, reasoned objections to competing views, and other devices familiar to philosophers—it presumably will *not* include dogmatic appeals to divine revelation, historical authorities, or prevailing social norms. Thus, we may say that *philosophy* as a way of life involves both accepting (1) some conception of the good life and engaging in (2) spiritual exercises to help one lead the specified good life, where (3) these tasks are undertaken in a rationally reflective fashion. In other words, to pursue philosophy as a way of life is to accept some conception of the good life on the basis of rational reflection *and* to engage in spiritual exercises that, again on the basis of rational reflection, one takes to be conducive to the accepted conception of the good life. Plausibly, this combination provides tools for bridging the theory–action gap, for (2) provides techniques for living in accordance with (1). Most ethical theories defend some conception of the good life but have nothing to say about how it might be achieved in practical terms.

Rational reflection regarding (1) is familiar, since contemporary value theorists continue to provide arguments for and against certain theories of the good life, such as hedonistic, preference–satisfaction, and objective list theories. The Hellenistic schools of philosophy largely shared the view that the good or flourishing life consisted of *ataraxia*—roughly, freedom from disturbance or stress—although they differed on what constituted *ataraxia* and how it was to be achieved (Nussbaum 1994, 41). For example, the Epicureans held that the good life consisted of freedom from pain, a state most reliably acquired by focusing on only natural and necessary pleasures (Long and Sedley 1987, 113–114). Alternatively, the Stoics advocated living according to reason, which they took to require subduing one's passions and freeing oneself from their disturbances (Nussbaum 1994, 316–358).

We can evaluate different conceptions of the good life in terms of whether and how well they deliver on the goal of *ataraxia*, such as by assessing the competing arguments offered by the Epicureans and Stoics for their respective views. But we also can evaluate how conceptions of the good life as *ataraxia* compare to other conceptions, such as by considering objections to Epicurean and Stoic positions. A life devoted to the pursuit of *ataraxia* might be incompatible with pursuits that *prima facie* have great value, such as cultivating personal relationships. Since there is always a risk of substantial disturbance in such relationships (e.g., due to a friend's betrayal), it might be difficult to see how pursuing *ataraxia* would be compatible with maintaining personal relationships. Yet it would be deeply counter-intuitive to hold that personal relationships have no place in a flourishing life. So we might conclude that the Stoics and Epicureans put too much emphasis on freedom from disturbance as the end of the good life (Nussbaum 1994, 9).

Initially, it might be less clear what would count as a rational reflection regarding spiritual exercises. A useful example of this is provided by Stoic and Epicurean disagreement regarding the *praemeditatio malorum*, or the practice of meditating on future ills, such as death. Both schools agreed that a person leading the good life would not be troubled by the prospect of her own death, but they diverged on whether meditation on death was a valuable exercise. According to Cicero, Epicurus suggested that one should not engage in the *praemeditatio malorum* in general, since doing so is apt to cause unnecessary disturbance (i.e., pain), thus inhibiting one's enjoyment of the good life. Rather than distressing oneself by considering future ills—some of which, death excluded, are not inevitable anyway—one should meditate on pleasure, including the past pleasures that one has enjoyed (Cicero 1927, 3.15.32–33; Foucault 2005, 468–469). Unlike meditating on painful prospects, meditating on pleasurable experiences in one's past can itself bring new pleasure in this very act of recalling the past. For the Epicurean, this spiritual exercise is therefore to be preferred to that of the *praemeditatio malorum*, because the former is far more effective at achieving the freedom from pain that constitutes one's flourishing.

The Stoics, conversely, suggest that meditating on death is a useful exercise because it can mitigate or remove anxiety from which we might otherwise suffer. For example, Marcus Aurelius holds that we should engage in meditative practices that see death as simply a part of the natural order:

> at all times awaiting death with contented mind as being only the release of the elements of which every creature is composed. If it is nothing fearful for the elements themselves that one should continually change into another, why should anyone look with suspicion upon the change and dissolution of all things? For this is in accord with nature, and nothing evil is in accord with nature.
>
> (Aurelius 1983, p. 2.17)

Here the *praemeditatio malorum* is thought to remove certain misconceptions and instill a proper understanding of death, namely that it is part of a rationally ordered nature and therefore not an ill after all. If we do not meditate on death, there is a risk both that we shall retain the mistaken view that death is evil and that we shall suffer distress from this mistaken view.

Both Epicurus and Marcus Aurelius offer competing arguments for their divergent views on the value of the *praemeditatio malorum*. To assess these arguments is an instance of rational reflection on (2). We may ask which argument is stronger, whether either is subject to damaging objections, whether the arguments rely on implausible premises, and so on. In this case, we can examine whether the *praemeditatio malorum* is successful in mitigating anxiety about one's own death, considering Stoic and Epicurean arguments for and against the practice. Of course, in making their arguments, both Marcus Aurelius and Epicurus appeal to doctrines of the Stoic and Epicurean schools, respectively. Thus, the task of rationally evaluating (2) is not here entirely separate from either the task of rationally evaluating (1) or the task of rationally evaluating other philosophical doctrines that may be relevant. For example, Marcus Aurelius relies on the Stoic idea that nature is rationally ordered, a view that is rejected by the atomistic Epicureans. Thus, in evaluating the Stoic advancement of the *praemeditatio malorum*, we may need to investigate whether something like the Stoic doctrine is true and, if not, what difference this would make for the value of meditating on future ills as a spiritual exercise. But this does not alter the fact that we can evaluate the effectiveness of the spiritual exercises themselves, even if that evaluation does not occur in isolation.

Finally, we should distinguish the study of philosophy as a way of life from the practice of philosophy as a way of life. As an example of the former, a scholar might be interested in studying (1)–(3) in the case of the Stoics but have no interest in incorporating (1)–(3) in her own life. Conversely, the Stoics themselves put (1)–(3) into practice in their own lives. While I assume that the study of philosophy as a way of life is a valuable enterprise, in the end, I am interested in the practice of philosophy as a way of life. As will become clear, my position is that environmental philosophy offers an attractive set of possibilities for this practice.

We are now in a position to sketch what counts as environmental philosophy as a way of life. If philosophy as a way of life consists of (1) accepting a conception of the good life and (2) adopting spiritual exercises that help one lead the good life, where (3) both these activities are pursued in a rationally reflective manner, then *environmental* philosophy as a way of life consists of (1') accepting an environmental conception of the good life and (2') adopting a set of environmental spiritual exercises meant to cultivate that good life, where (3) both these ends are pursued in a rationally reflective manner.[1] This formulation is appropriately open-ended, given the many forms of (1') and (2') that seem possible. I discuss the prospects for (1') and (2') later in this chapter.

Virtue and Flourishing

It is no accident that practitioners of philosophy as a way of life tend to emphasize the virtues. First, virtue plays an important role in various conceptions of the good life. Second, many spiritual exercises are designed to cultivate virtue and extirpate vice. Viewed as an excellent character trait, virtue is sometimes taken to be at least partly constitutive of an individual's own flourishing (Hursthouse 1999).[2] Some ancient philosophers held that virtue was even sufficient to secure a flourishing life. Plotinus, for example, holds that the genuine sage (i.e., the fully virtuous person) could maintain his connection to the Good even while being tortured to death, and Epicurus held that the sage would find such torture pleasant (Cicero 1927, 2.7.17–18; Plotinus 1969, 1.4.13). Alternatively, Aristotle held that virtue was necessary but not sufficient to secure a flourishing life. Since factors outside of one's control (e.g., poor health) can negatively impact one's flourishing, even being fully virtuous is not a guarantee that one will lead the good life. Accordingly, Aristotle holds that those who insist that one can flourish while being tortured to death are mistaken (Aristotle 1999, 1153b). Nonetheless, even on this less radical Aristotelian view, being a virtuous individual is required in order to flourish or achieve the good life, inasmuch as virtue is at least part of what constitutes such flourishing.[3]

In fact, Aristotle also argues that certain virtues, such as the greatness of the soul, enable one to bear misfortune well (Aristotle 1999, 1124a). While this does not cancel his claim that external factors play some role in whether or not one is flourishing, it does suggest that being virtuous can reduce the impact such factors can have on our flourishing, although not to the extent supposed by Epicurus and Plotinus.

An interest in virtue is natural for any philosophy concerned with how to lead the good life. At least in part, this is because leading a good life is not plausibly construed as a series of discrete actions, as if living well consisted of merely performing a sequence of disconnected acts. It is doubtful whether such a series would count as a life at all, much less a life that is properly *led*. Rather, leading a life seems to require integration among one's various actions, an integration rooted in one's character. For if one's actions were not rooted in some character, it is difficult to see how those actions would have the unity or coherence that seems necessary to constitute a genuine life. To put matters simplistically for the moment, one's character is constituted by certain traits, some of which may be good (virtues) and some of which may be bad (vices). Leading a *good* life might be taken to require in part that one have a good character or one constituted by virtues. At any rate, since this was the view of many Hellenistic philosophers, it is not surprising that many of them emphasized the importance of virtue since being virtuous is arguably necessary for the good life they hoped to pursue.

Further, in attempting to lead the good life, looking to virtuous exemplars seems helpful (Hursthouse 1996). This is especially so given that developing virtuous character traits may be tied up with internalizing principles of action, forming certain habits, or modifying one's dispositions and desires. Here an abstract rule or decision procedure regarding how to act is likely insufficient, and so it may be useful to consider actual virtuous individuals, learning from the qualities they display, how they themselves became virtuous (e.g., through some spiritual exercises they might perform), the activities they avoid, the motivations they report having, and the advice they might offer. Indeed, if they are able to communicate with us, such exemplars also can serve as teachers of the good life. Especially when it comes to pursuing (2), it is no doubt extremely difficult to train oneself without aid. More experienced practitioners can offer guidance and support, corrections to various mistakes (e.g., faulty techniques), and so on. One sees this in the letters of Seneca to Lucilius, in which the teacher offers his pupil advice, encouragement, and correction as Lucilius attempts to lead the Stoic life (Seneca 1961; Foucault 1986, 53).

Finally, I should note that placing an emphasis on virtue as a constituent of the good life does not require one to adopt a virtue ethic as a matter of normative ethical theory, although it is compatible with doing so. That is, one need not be a virtue *ethicist* in order to recognize the importance of virtue for the good life. One reason for this is that to pursue (1)–(3) is not to commit oneself to a normative ethical theory that might serve as a competitor to other such theories. To see why this is so, note that pursuing (1) and (2) is compatible with accepting either a consequentialist or a deontological normative ethic. While proponents of both these kinds of theory do allow an important role for virtue (Nussbaum 1999), neither consequentialists nor deontologists are virtue ethicists in the proper sense. Yet it is perfectly conceivable for someone both to pursue philosophy as a way of life by cultivating certain virtues and to follow the dictates of some version of consequentialism (e.g., by maximizing happiness) or deontology (e.g., by always respecting persons as ends-in-themselves). On non-virtue ethical theories, there is at least nothing wrong about generally pursuing (1) and (2) in addition to satisfying one's moral obligations.[4] Further, some non-virtue ethicists coherently maintain that we have a moral obligation to develop virtues. Kant, for example, holds that one has a moral duty to herself to cultivate virtuous dispositions (Kant 1999, 6:446; Svoboda 2012). Yet Kant is not a virtue ethicist, since this duty fits within a broader deontological framework.

Why Philosophy as a Way of Life Is Worth Reviving

I take it as a truism that any human person has good reason to lead a good life and that many of us in fact desire to do so. We can assume broad agreement on this general point. Controversy arises when some specific conception of the good life is put forward. Not surprisingly, there is disagreement on a host of issues that seem relevant here, such as the role of pleasure in the good life and the contribution of virtue to one's own flourishing. Because of this disagreement, becoming clear on available accounts of the good life, thinking about their respective merits and deficiencies, and adopting some such conception—effectively pursuing (1)—is worthwhile. But if the pursuits of thinking about the good life and accepting some conception of it are worthwhile, then surely it is also worthwhile to consider and implement means by which we might succeed in living according to that conception. This would involve the development and practice of certain exercises meant to cultivate oneself in such a way that leading the specified good life becomes possible or more manageable. Effectively, this is to pursue (2). Lacking this, we would have failed to bridge the theory–action gap. That is, we might have a sophisticated, plausible, and well-argued theory of the good life, but this by itself would provide us no help in actually living well. We need means to become the sorts of persons specified by the theory in question, to put that theory into practice. Spiritual exercises played precisely this role in the philosophical traditions I have mentioned.

The foregoing considerations are not sufficient to establish that *philosophy* as a way of life is worth reviving, for various non-philosophical approaches are available in pursuing both (1) and (2). One might be impressed by some organized religion's conception of the good life and commence the ritual or meditative practices associated with it, or one might read self-help books and act on their advice. If we are to believe that a distinctively philosophical approach to (1) and (2) is worth reviving, then we need some plausible reason to think that a rationally reflective approach to (1) and (2) is worth pursuing. Importantly, providing a compelling reason to accept this need not entail that non-philosophical approaches to (1) and (2) are inferior to a philosophical one. *Perhaps* philosophy as a way of life is preferable to these other approaches, but I will not argue for that claim here. Instead, I suggest that philosophy as a way of life is worth reviving at least as an additional option to these other approaches. There are two quite plausible reasons to accept this weaker claim.

First, a philosophical approach to (1) and (2) should be attractive to those who already put significant value on rational reflection. Given that the practice of philosophy as a way of life involves subjecting both conceptions of the good life and suggested spiritual exercises to close rational scrutiny, those who are skeptical of available accounts of the good life may find this philosophical approach attractive. Philosophy as a way of life encourages its practitioners to evaluate and question competing approaches to (1) and (2). There are a variety of reasons why, on reflection, one might rationally reject some candidate for the good life—perhaps it is internally incoherent, incompatible with our best science, morally indefensible, dependent upon fantastical historical claims or dubious appeals to divine revelation, and so on. For those who value such reflection, the practice of philosophy as a way of life would seem to offer a valuable approach to thinking about the good life and ways to lead it.

Second, a philosophical approach to (1) and (2) offers ways to navigate the controversy and disagreement that are inevitable when it comes to questions regarding competing conceptions of the good life and associated spiritual exercises. While it would be naïve to think that philosophical argumentation will produce a single consensus on what the good life is and what spiritual exercises will help us lead it, rational reflection nonetheless allows for reasoned dialogue among defenders of competing views. If we are uncertain what the good life is, we can consider the relative merits of competing accounts, as well as objections that have been raised against them. One might thereby come to a justified (or at least reasonable) position regarding (1) and (2), be able to defend that position with plausible arguments, and be able to critique competing positions. Lacking such argumentative tools, we might be unable to say why some way of life is preferable to others, and this might leave us at a loss regarding which way of life to adopt and/or which spiritual exercises to undertake. Given its commitment to rational reflection, the practice of philosophy as a way of life is attractive in part because it offers a non-dogmatic way to navigate widespread disagreement regarding how to live well.

Environmental Virtue and Conceptions of the Good Life

In the remainder of this chapter, I argue that environmental philosophy has the resources to renew the practice of philosophy as a way of life and that this is a worthwhile goal. In the present and subsequent sections, I respectively suggest that there is substantial material in environmental philosophy for developing plausible versions of (1') environmental conceptions of the good life and (2') environmental spiritual exercises meant to realize or at least approximate such conceptions. Although I have developed neither a full-fledged conception of the environmental good life nor a worked-out set of environmental spiritual exercises, this discussion will indicate some plausible and attractive forms environmental philosophy as a way of life might take.

Environmental philosophers recently have placed a great deal of emphasis on environmental virtues and vices (Cafaro and Sandler 2005; Sandler 2009). Among environmental virtues, we might count benevolence toward non-human animals, humility regarding one's place in the natural world, respect for nature, and temperance in one's use of natural resources. Among environmental vices, we might count malevolence toward animals, an arrogant attitude of human superiority, and greedy exploitation of natural resources. While many of those who write on environmental virtue and vice accept some kind of environmental virtue ethic as a normative theory, an emphasis on environmental virtue in one's thinking does not require one to adopt such a theory (Svoboda 2015). For example, Paul Taylor makes much of the virtue of respect for nature, yet he develops a broadly deontological framework for moral obligation to biotic entities (Taylor 1986, 198–218). Without defending the content of his account, this example suggests that recognizing the importance of environmental virtues does not require accepting an environmental virtue ethic *per se.*

It is an open question whether or not environmental virtues belong to a distinct class of virtue. On the "extensionist" view, environmental virtues (and vices, *mutatis mutandis*) are simply non-environmental virtues extended to cover environmental cases. On the "non-extensionist" view, at least some environmental virtues belong to a distinct class—they are essentially environmental in character and thus do not consist merely of extending non-environmental virtues (Sandler 2005, 219–220). I will not take a position here on this controversial issue. Fortunately, it is not necessary for me to take a position on this matter, because both positions share the view that there are genuine environmental virtues and vices. If nothing else, we at least should accept the extensionist claim that commonly recognized virtues and vices sometimes cover environmental cases. Surely it is possible to be benevolent toward non-human animals or greedy in one's use of natural resources, for example, even if the benevolence or greed at issue is no different in kind from that involved in non-environmental cases. Since I take it that either extensionism or non-extensionism is true, and since on either view there are genuine environmental virtues, I conclude that there are genuine environmental virtues.

For this reason, it is plausible to suppose that one can be environmentally virtuous. Now if we accept the view that virtue is necessary (if not sufficient) for flourishing, then we have some initial reason to suspect that possessing environmental virtues could contribute to a flourishing life.[5] Upon reflection, it seems reasonable to identify an environmentally virtuous person as flourishing to a greater degree than an environmentally vicious person, all else being equal. So-called "last person" scenarios—in which the last person on earth destroys entire ecosystems for amusement, but without any possibility of his actions affecting present or future humans— make this explicit. The last person's actions seem not only to be morally wrong (Sylvan 2003) but also to indicate the presence of bad character traits (perhaps arrogance or malevolence) that seem inimical to the last person's own flourishing (O'Neill 1992). We might ask what sort of *person* would engage in such horrific actions (Hill 2005). The last person contrasts with an environmentally virtuous person, say one who displays the traits of benevolence toward non-humans and respect for nature. We might think that all else being equal, the environmentally virtuous person is flourishing to a greater extent than the last person. This judgment is well-explained by the notion that virtue is constitutive of flourishing or that the virtues benefit their possessor (Hursthouse 1999). This is not primarily a point about the *moral* badness of the last person's character and the *moral* goodness of the virtuous person's character. Rather, the point is that the last person seems to fail in the task of leading a good life, and this is harmful to himself. Alternatively, the environmentally virtuous person seems to avoid such harm to herself, arguably because she is benefited by her own virtues.

In addition to a rich literature on environmental virtue, there is also no shortage of environmentally virtuous exemplars to whom we might look, such as Rachel Carson, Aldo Leopold, John Muir—and Henry David Thoreau is an obvious case (Cafaro 2005). We may plausibly take Thoreau to have pursued philosophy as a way of life (Hadot 2005). He famously announces in *Walden*,

> I went to the woods because I wished to live deliberately, to front only the essential facts of life, and see if I could not learn what it had to teach, and not, when I came to die, discover that I had not lived.
>
> (Thoreau 2004, 88)

Much of that book is an illustration of Thoreau's commitment to environmental virtues like simplicity and respect for nature. Moreover, Thoreau seems to have valued such virtues for the contribution they can make to a flourishing life. In his "Life Without Principle," he excoriates his fellow citizens for being more concerned with "business" and material acquisition than with living well, suggesting that wisdom essentially involves both knowing how to lead the good life and choosing to pursue it (Thoreau 2013). Indeed, Thoreau arguably practiced *environmental* philosophy as a way of life, conceiving of the good life as constituted in part by environmental virtues, as well as engaging in various spiritual exercises (see below) meant to instill and reinforce them.

To those who share something close to Thoreau's conception of the good life, he seems to offer a compelling exemplar. For those who do not accept Thoreau's conception of the good life, there are other environmentally virtuous exemplars to whom they might appeal. The existence of such exemplars provides further reasons to think that environmental philosophy as a way of life can get off the ground. First, it suggests that cultivating environmental virtue is feasible, given that others have already done so. Second, the pool of such exemplars provides a kind of resource to those who wish to lead an environmentally virtuous life. We can look to the lives of Carson, Leopold, Muir, and Thoreau in order to compare different conceptions of the environmentally virtuous life, and we can draw upon their experiences and counsel in order to avoid pitfalls and pursue avenues that are more likely to be successful in leading such a life.

In general, both a rich literature on environmental virtue and a history of environmentally virtuous individuals suggest that environmental philosophy has the resources to pursue (1'), developing a conception of the good life that (at least in part) includes the possession of environmental virtues. Developing a conception of (1') seems worthwhile, given the plausibility of the view that the environmentally virtuous person flourishes to a greater extent than the environmentally vicious person, all else being equal. Of course, divergent conceptions of (1') are possible. Evaluating different candidates for the environmental good life would require becoming clear on their relative merits and deficiencies through rational reflection. Environmental philosophers are already pursuing this task to some extent, particularly in the literature on environmental virtue.

Environmental Spiritual Exercises

Members of the Hellenistic schools practiced a variety of exercises meant to assist their pursuit of (1). Some of these exercises, such as keeping journals, functioned to internalize relevant beliefs and values. The best-known example of this is the *Meditations* of Marcus Aurelius, in which he urges himself to remember and live by the doctrines of the Stoic school and reflects on how to overcome disturbances in his soul. We might view this exercise of journal writing as helping to secure one's commitment to a certain conception of the good life, as well as sculpting the habits and desires conducive to that good life. The practice allows one to remind oneself of the values to which she is committed, to exhort herself to act in accordance with them, to reflect on how and why her recent behavior has failed (or succeeded) in approximating these values, and so on (Foucault 1997). This qualifies as a spiritual exercise, given that it aims to bring about a transformation in the practitioner, cultivating her habits and desires in accordance with some conception of the good life.

Another ancient spiritual exercise consisted of self-examination, in which one reflected on how she had lived during the previous day. Here one considers whether her actions and thoughts were in accordance with the relevant conception of the good life, identifying mistakes that might not have been obvious otherwise (Seneca 1979, 3.36). As Foucault notes, the point of this exercise was not to feel remorse but rather "to enhance the rational equipment that ensures a wise behavior" (Foucault 1986, 62). Suppose that one's conception of the good life involves the absence of the emotions of hatred, shame, and envy. Self-examination at the end of one's day would then include considering whether one harbored those emotions at any point since waking that morning. If one finds that she has experienced these emotions, further self-examination may diagnose their causes. Understanding these causes may provide the "rational equipment" that can help one avoid such emotions in the future, such as by making one realize that these emotions depend on our own interpretation of events or of the actions of other persons, interpretations over which we have some control. This spiritual exercise thus can render one more sensitive to obstacles to the good life, and it can equip one to overcome these obstacles.

For an *environmental* spiritual exercise, reconsider the case of Thoreau, who was committed to various practices conducive to a conception of the good life that places a high value on simplicity, acceptance of nature (Hadot 2005), and perhaps even *ataraxia*. These practices included walking excursions, voluntary poverty, and journal writing—but Thoreau's approach to manual labor is especially instructive. He held that the beans he planted at Walden would be partially consumed by wildlife and thus "grow for woodchucks partly." Rather than worrying about this, Thoreau suggests that the "true husbandman will cease from anxiety, as the squirrels manifest no concern whether the woods will bear chestnuts this year or not" (Thoreau 2004, 161). Thoreau's position is similar to a central thought of Epictetus: "Do not seek to have everything that happens happen as you wish, but wish for everything to happen as it actually does happen, and your life will be serene" (Epictetus 1928, 8). Attempting to make the world fit one's desires is likely to meet with disappointment, whereas fitting one's desires to the world can decrease anxiety. Thoreau seems to agree that some matters are not worth disquiet—it is sometimes best to acquiesce, such as by accepting nature's consumption of a portion of the product of one's labor. This attitude of "true husbandry" helped make Thoreau's manual labor into a spiritual exercise, a point previously urged by the Stoic Musonius Rufus, who held that manual labor can serve to teach and internalize philosophical lessons (Rufus 2011).

Thoreau's manual labor counts as an environmental spiritual exercise because it helped to instill acceptance of nature, which he deemed to be a component of the good life. This is a *spiritual* exercise because it is a practice meant to effect a transformation in the practitioner. It is an *environmental* spiritual exercise because it employs interaction with the natural world in order to craft or strengthen an attitude regarding nature. If successful in crafting or strengthening a component of the good life, such a practice helps one bridge the theory-action gap. It provides a practical way to achieve a component (e.g., acceptance of nature) of the good life. For this reason, environmental spiritual exercises offer resources for putting values and norms into practice. Unlike purely theoretical approaches, environmental philosophy as a way of life does not merely specify what constitutes the good life—it also provides the "rational equipment" (to borrow Foucault's phrase) for achieving such a life.

Spiritual exercises need not be thoroughly practical. They also can involve a theoretical component. Hadot argues that virtually all of the Greco-Roman schools of philosophy advocated "the view from above," a kind of vantage point on the world and on human affairs meant to contribute to one's progression toward *ataraxia*. For Epicureans, Stoics, Platonists, and even the Skeptics,

> philosophy was held to be an exercise consisting in learning to regard both society and the individuals who comprise it from the point of view of universality. This was accomplished partly with the help of a philosophical theory of nature, but above all through moral and existential exercises. The goal of such exercises was to help people free themselves from the desires and passions which troubled and harassed them.
>
> (Hadot 1995, 242)

As Hadot notes, adopting the view from above has both theoretical and practical dimensions. On the one hand, a "philosophical theory of nature"—such as the Stoic idea that nature is infused with a providential, cosmic reason (Hadot 1995)—may help one abstract from individual concerns, stress, pain, and so on. If one views the universe as a providentially governed cosmos, for example, it is perhaps easier to find meaning in one's suffering, seeing it as part of the whole and achieving some degree of equanimity with regard to it. On the other hand, adopting the view from above (e.g., through meditation) might itself serve as a spiritual exercise, offering means by which to overcome disturbances, perhaps by making it easier to view them from a third-person perspective. In effect, this can help one regard her own situation from "the point of view of universality," rather than from a first-person perspective in which such disturbances are experienced immediately. Importantly, the view from above is not equivalent to "the view from nowhere," the latter of which may be impossible to adopt for human beings. The view from above is still a view from *somewhere*, but from a place that affords some distance in observing the phenomenon in question. For example, one might view one's own suffering as if it were the suffering of someone else. This still involves a perspective, namely a third-person one, but it allows for some degree of detachment from the suffering one might otherwise experience. Unlike the view from nowhere, the view from above seems possible to adopt, such as through an act of imagination.

Like the ancient schools, environmental philosophy includes various theories that seem consonant with the view from above. Many environmental philosophers urge us to recognize the intrinsic value of non-human nature and the consequent moral standing of non-human entities (Rolston 1982), and many have emphasized future generations and our obligations to them, requiring us to consider how present actions may impact even the distant future (Shrader-Frechette 2000). Arguably, these theoretical positions encourage something close to the view from above, because adopting them expands our vantage point beyond merely individual, present, or human concerns. These environmental philosophical views may have a practical dimension relevant to one's own flourishing. Like the Stoics' cosmic reason, seeing ourselves as only part of an intrinsically valuable nature, or as but one of many equally important generations, may make it easier to put our own disturbances, stress, and suffering in perspective, viewing it as part of some larger sequence, collection, or whole. *Prima facie*, these views seem to offer theoretical resources for dealing with threats to our own *ataraxia*, helping us to abstract from a first-person perspective on our own disturbances and instead adopting "the point of view of universality" with respect to them. This also may help us cultivate environmental virtues, such as acceptance of nature or benevolence toward intrinsically valuable non-humans. If all this is correct, then such theories in environmental philosophy can be understood as making contributions to (2'), since internalizing these theoretical views can help one cultivate her life in accordance with (1'), or some environmental conception of the good life. One might, for example, repeatedly meditate on the vastness of nature, which is to adopt a version of the view from above. Over time, this meditative practice has the potential to effect a transformation in oneself, building environmental humility or acceptance of nature, for example. Plausibly, this practice helps to remove obstacles to attaining these environmental virtues, such as fixation on disturbances in one's life. By routinely adopting this environmental form of the view from above, one can come to see disturbances in one's life as miniscule parts of an enormous natural world. In that case, we can take this form of meditation to be an environmental spiritual exercise, for it instills environmental virtues through a transformative practice.

Yet one might question whether the view from above is actually a good perspective to adopt. First, perpetually maintaining this view presumably would be impossible for human beings. It is difficult even to imagine someone who never regards her own disturbances from a first-person perspective. Second, adopting the view from above might involve a problematic disengagement, perhaps by creating a kind of affective detachment or even apathy regarding, for example, the suffering of other persons. However, advocating the view from above need not involve a directive to maintain this view at all times, even if some ancient sources seem to advise doing so. Rather, adopting the view from above may be suggested as a temporary exercise meant to assist one's pursuit of the good life, given that *sometimes* we can be too invested in things closest to us. For example, in the grip of intense or prolonged suffering, it often may be both helpful and appropriate to attempt to view that suffering and its causes from a universal standpoint. This does not require us to adopt an apathetic stance toward others, since we may still hold that there are times when it is appropriate to adopt a first-person perspective. For example, it may not be appropriate to extirpate all grief regarding the death of a friend, but there may be times when intense and prolonged grief should be assuaged, and the view from above may help do so. While precisely when the view from above is appropriate would depend on the specific conception of (1') in question, it would be too hasty to deny that this perspective can be useful.

Which environmental spiritual exercises should be practiced will depend on the conception of (1') that is adopted. If some conception of the environmental good life grants an important place to *ataraxia*, then candidates for (2') can be evaluated partly for how well they assist their practitioners in approximating a life free of disturbance. Of course, environmental philosophy as a way of life need not take *ataraxia* to be a constituent of the good life. Perhaps instead of (or in addition to) *ataraxia*, a conception of (1') should include possessing certain environmental virtues. Fortunately, environmental philosophy offers remarkably rich possibilities for distinctively environmental spiritual exercises, ones that are compatible with various conceptions of (1'). I have discussed two options already: meditation on the vastness of nature and, by way of the example of Thoreau, interaction with nature through manual labor.

Another set of environmental spiritual exercises is tied to our dietary practices. Adopting a vegan or vegetarian diet in a certain way could, for example, habituate one to prioritizing the well-being of non-human animals over one's immediate desires (e.g., to enjoy the taste of meat), thereby cultivating environmental benevolence. In order to count as a spiritual exercise, such a practice must involve more than merely avoiding meat or animal products. In addition to that, one will remain mindful of why she eats certain foods and avoids others, and so her daily practices can serve as reminders of the environmental values she seeks to honor and the environmental virtues she seeks to craft or maintain. Another type of environmental spiritual exercise is environmental writing. For example, recording one's environmental experiences in writing, as Leopold does in *A Sand County Almanac*, might serve as a transformative practice, perhaps helping to internalize and reinforce the values and beliefs that contribute to the pursuit of certain environmental virtues, like respect for nature (Leopold 2001). The act of writing allows one to recall, appreciate, and organize one's experiences in the natural world. It can serve as a reminder of the value of those experiences.

I have argued both that philosophy as a way of life is worth reviving and that environmental philosophy is well-suited to enact such a revival. Environmental philosophy as a way of life would involve both (1') some distinctively environmental conception of the good life and (2') a set of environmental spiritual exercises to cultivate that life in oneself. I have suggested some possible forms (1') and (2') might take, but I have neither advocated nor argued for any particular conception of the environmental good life, nor for any particular set of environmental spiritual exercises. In the next chapter, I will attempt to show how and why this general framework is attractive for ecological pessimists.

Notes

1 I do not suggest a (3'), because I see no reason to suspect that the rational reflection involved would or should be any different in kind from that involved in non-environmental forms of philosophy as a way of life.

2 I will not attempt the difficult task of determining precisely what a virtue is. While this is an interesting question, the more general conception of virtue as a good character trait is sufficient to allow discussion of what role virtue plays in the good life.

3 Importantly, virtue need not be limited to *morally* excellent character traits alone. We can recognize a wide range of excellent character traits, including intellectual and physical virtues, for example. It may be that the good life includes both moral and non-moral virtues.

4 Of course, there might be specific cases in which pursuing (1) and (2) at a particular time would conflict with some other moral obligation one has, such as when one has a duty to interrupt some spiritual exercise in order to assist someone in an emergency situation. But cases like this do not threaten the general point.

5 It is possible that some but not all virtues contribute to flourishing, and it is also possible that environmental virtues are among those that do not so contribute, but I see no reason to suppose that environmental virtues would be exceptional in this way.

References

Aristotle. *Nicomachean Ethics*. Translated by T. Irwin. Indianapolis: Hackett, 1999.

Aurelius, M. *The Meditations*. Translated by G. M. A. Grube. Indianapolis: Hackett, 1983.

Booth, C. "A Motivational Turn for Environmental Ethics." *Ethics & the Environment* 14, no. 1 (1999): 53–78.

Cafaro, P. "Thoreau, Leopold, and Carson: Toward an Environmental Virtue Ethics." In *Environmental Virtue Ethics*, edited by P. Cafaro and R. Sandler. Lanham: Rowman and Littlefield, 2005.

Cafaro, P. and R. Sandler, eds. *Environmental Virtue Ethics*. Lanham: Rowman and Littlefield, 2005.

Cicero. *Tusculan Disputations*. Translated by J. E. King. Cambridge: Harvard University Press, 1927.

Coeckelbergh, M. *Environmental Skill: Motivation, Knowledge, and the Possibility of a Non-Romantic Environmental Ethics*. New York: Routledge, 2015.

Davidson, A. "Introduction." In *Philosophy as a Way of Life: Spiritual Exercises from Socrates to Foucault*, edited by P. Hadot, translated by M. Chase. Hoboken: Wiley, 1995.

Epictetus. *Epictetus: Discourses, Books 3–4; The Encheiridion*. Cambridge: Harvard University Press, 1928.

Foucault, M. *The Care of the Self: Volume 3 of the History of Sexuality*. Translated by R. Hurley. New York: Random House, 1986.

Foucault, M. "Self-Writing." In *Ethics: Subjectivity and Truth*, edited by P. Rabinow. New York: New Press, 1997.

Foucault, M. *The Hermeneutics of the Subject: Lectures at the Collège de France 1981–1982*. Translated by G. Burchell. New York: St. Martin's Press, 2005.

Goralnik, L. and M. P. Nelson. "Framing a Philosophy of Environmental Action: Aldo Leopold, John Muir, and the Importance of Community." *The Journal of Environmental Education* 42, no. 3 (2011): 181–192.

Hadot, P. *Philosophy as a Way of Life: Spiritual Exercises from Socrates to Foucault*. Translated by M. Chase. Hoboken: Wiley, 1995.

Hadot, P. *What Is Ancient Philosophy?* Translated by M. Chase. Cambridge: Harvard University Press, 2002.

Hadot, P. "There Are Nowadays Professors of Philosophy, But Not Philosophers." *Journal of Speculative Philosophy* 19, no. 3 (2005): 229–237.

Hill, T. "Ideals of Human Excellence and Preserving Natural Environments." In *Environmental Virtue Ethics*, edited by P. Cafaro and R. Sandler. Lanham: Rowman and Littlefield, 2005.

Hursthouse, R. "Normative Virtue Ethics." In *How Should One Live?*, edited by R. Crisp. Oxford: Oxford University Press, 1996.

Hursthouse, R. *On Virtue Ethics*. Oxford: Oxford University Press, 1999.

Irvine, W. *A Guide to the Good Life: The Ancient Art of Stoic Joy*. New York: Oxford University Press, 2008.

Kant, I. *Practical Philosophy*. Translated by M. J. Gregor. Cambridge: Cambridge University Press, 1999.

Kretz, L. "Climate Change: Bridging the Theory-action Gap." *Ethics & The Environment* 17, no. 2 (2012): 9–27.

Leopold, A. *A Sand County Almanac*. New York: Oxford University Press, 2001.

Long, A. A. and D. N. Sedley. *The Hellenistic Philosophers: Volume 1, Translations of the Principal Sources with Philosophical Commentary*. Cambridge: Cambridge University Press, 1987.

Nussbaum, M. *The Therapy of Desire: Theory and Practice in Hellenistic Ethics*. Princeton: Princeton University Press, 1994.

Nussbaum, M. "Virtue Ethics: A Misleading Category?" *The Journal of Ethics* 3, no. 3 (1999): 163–201.

O'Neill, J. "The Varieties of Intrinsic Value." *Monist* 75-, no. 2 (1992): 119–137.

Plotinus. 1969. *Plotinus: Porphyry on Plotinus, Ennead I*. Translated by A. H. Armstrong. Cambridge: Harvard University Press.

Rolston III, H. "Are Values in Nature Subjective or Objective?" *Environmental Ethics* 4, no. 2 (1982): 125–151.

Rufus, M. *Musonius Rufus: Lectures and Sayings*. Translated by C. King. Scotts Valley, CA: CreateSpace, 2011.

Sandler, R. "A Virtue Ethics Perspective on Genetically Modified Crops." In *Environmental Virtue Ethics*, edited by P. Cafaro and R. Sandler. Lanham: Rowman and Littlefield, 2005.

Sandler, R. *Character and Environment: A Virtue-Oriented Approach to Environmental Ethics*. New York: Columbia University Press, 2009.

Seneca. *Epistulae Morales, Volume I*. Translated by R. M. Gummere. Cambridge: Harvard University Press, 1961.

Seneca. "De Ira." In *Seneca, Moral Essays, Volume 1*. Translated by John W. Basoe. Cambridge: Harvard University Press, 1979.

Shrader-Frechette, K. "Duties to Future Generations, Proxy Consent, Intra- and Intergenerational Equity: The Case of Nuclear Waste." *Risk Analysis* 20, no. 6 (2000): 771–778.

Svoboda, T. "Duties Regarding Nature: A Kantian Approach to Environmental Ethics." *The Kant Yearbook* 4 (2012): 143–163.

Svoboda, T. *Duties Regarding Nature: A Kantian Environmental Ethic*. New York: Routledge, 2015.

Sylvan, R. "Is There a Need for a New, an Environmental Ethic?" In *Environmental Ethics: An Anthology*, edited by A. Light and H. Rolston III. Malden: Blackwell, 2003.

Taylor, P. *Respect for Nature: A Theory of Environmental Ethics*. Princeton: Princeton University Press, 1986.

Thoreau, H. D. *Walden: A Fully Annotated Edition*. New Haven: Yale University Press, 2004.

Thoreau, H. D. "Life Without Principle." In *Essays: A Fully Annotated Edition*, edited by J. S. Cramer. New Haven: Yale University Press, 2013.

6 A Case for Meliorism

I will now argue that ecological pessimists should be meliorists. An important component of classical pragmatism, meliorism is simply the view that the world can be improved but not perfected. This is a sort of anti-utopian view. Something approaching a perfect world—perfectly just or happy, for example—may be unattainable, but we nonetheless can make some progress in that direction, perhaps a great deal of progress.

Meliorism in Classical Pragmatism

Interestingly, the pragmatists directly tie meliorism to issues of optimism and pessimism (see Liszka 2021). According to Charles Peirce, meliorism is the view "that the world is neither the worst nor the best possible, but that it is capable of improvement: a mean between theoretical pessimism and optimism" (Peirce 1889–1991, 3697). William James adds that meliorism "holds up improvement as at least possible" (James 1907, 119). For John Dewey, meliorism is "the belief that the specific conditions which exist at one moment, be they comparatively bad or comparatively good, in any event may be bettered" (Dewey 2008, 181–2; Peirce 1931–58, 2.181–2). Now Peirce defines meliorism in a manner seemingly at odds with my own position, viewing it as a sort of middle road between optimism and pessimism. However, Peirce associates pessimism with thinking this world to be the worst possible. That is not what I mean by the term. One can be pessimistic in the sense of merely expecting things to get worse. In my view, we are not doomed to the worst possible end, just likely a bad one. In effect I agree with Peirce: meliorism is "a mean" between extreme pessimism and optimism.

One way to improve the state of affairs is by mitigating ills, and one way of mitigating ills is to reduce ecological risk. We may think of this as the "negative" dimension of meliorism, as it involves negating bad phenomena: the magnitude and frequency of suffering, the extent of injustice, and so on. John Nolt estimates that the average American will be responsible

DOI: 10.4324/9781003570523-7

for "one or two" deaths in the future due to their emissions (Nolt 2011; see also Nolt 2015). That is bad. If we could reduce that average to .75, that would be less bad. Another application of meliorism is simply to maintain the status quo. Suppose that, without intervention, the average American's emissions increase to the point that they are responsible for four future deaths. We might enact certain policies (e.g., a carbon tax) preventing that increase, such that average emissions remain flat. This would count as an improvement relative to a counterfactual scenario in which we allowed emissions to increase. That too would be a form of meliorism.

To be clear, I am not advocating pragmatism in a global sense, but only borrowing the pragmatists' melioristic ideas. This is for two reasons. First, even limiting ourselves to the classic pragmatists, there is much diversity in their thought. There is not a single, monolithic form of pragmatism one might adopt, but rather many different varieties. Accordingly, my taking a pragmatic approach would require laying out and defending a particular variety of pragmatism. Second, there is much about classical pragmatism that is controversial, as I discuss below. It would be unfortunate (for me) if my approach stood or fell with controversial views on these other issues. So why discuss the classical pragmatist at all? Because they have made the best and most detailed case for meliorism, which is very helpful for the pessimist who wishes to avoid certain pitfalls. In short, I will remain agnostic about the various commitments of pragmatism, such that my view will be compatible with a wide range of metaphysical and epistemological positions. With that said, later in this chapter, I do sketch a generally "pragmatic" variety of philosophy as a way of life, but this should not be taken as an endorsement of pragmatism in other areas, such as epistemology.

This approach immediately raises a question. Is it possible to help oneself to pragmatic meliorism while overlooking the many other dimensions of pragmatist philosophy? Yes. It is certainly possible to be a meliorist without being a pragmatist in other respects. At the core of meliorism is the simple and reasonable idea that the world can be improved. This is a social-political-moral view. Obviously, one can believe this without being committed to substantive views in metaphysics or epistemology. For example, many of the pragmatists were metaphysical naturalists, holding (roughly) that reality is limited to spatio-temporal phenomena (Kim 2003). Yet it is clear that an anti-naturalist could also believe that the world is susceptible to improvement. In terms of metaphysics, meliorism commits us to little more than the view that change is possible. Whether that change is a matter of purely natural phenomena, ideas in the mind of God, or something else is a further, distinct question. Similarly, meliorism does not require any specific stance in epistemology. Many of the classical pragmatists were anti-foundationalists, denying that beliefs are ultimately justified by virtue of some basic, foundational set of beliefs (Margolis 1984). Once again, a meliorist need not be an anti-foundationalist. I can hold that there are basic, axiomatic truths that are known *a priori* while also holding that the world may be improved. Meliorism implies that we can have knowledge about what counts as an improvement, but that is a minimal commitment that fits with many epistemological theories.

The Compatibility of Pessimism and Meliorism

It is a mistake to suppose that pessimism and progress are incompatible. This mistake perhaps explains why some think that pessimism makes activism pointless. The idea might be that if we are doomed to a bad future, then there is no reason to work for something better (or less bad). This is just false. The mistake here is to forget what we saw in Chapter 4, namely that a bad future need not be the worst of all possible worlds. In my view, the future will be bad, but just how bad is yet to be determined. That this is the worst possible world, or that attempts at progress are futile (see Beiser 2016), belong to extremely pessimistic views of the sort I do not accept. To be explicit, I am not claiming that progress is merely possible. That would be an almost trivial point, depending on the type of possibility one has in mind. Avoiding ecological catastrophe is *possible*, but it is unlikely.

We need to make a distinction between what is merely possible in any given sense (e.g., logically, physically, politically) and what is feasible. As with mere possibility, there are various types of feasibility: political, social, technical, and so on. As I understand it, feasibility covers what is realistically doable given actual conditions. As with many other concepts I employ in this book, there is vagueness in what counts as "realistically doable," and there will certainly be room for reasonable disagreement about some cases, but I trust that many cases are obvious. For instance, ceasing all global greenhouse gas emissions within the next three months is not feasible. That would require drastic changes to the global economy and has no chance of happening. It is not realistically doable. But other measures are realistically doable, such as tax policies that incentivize the use of renewable energy. From an environmental point of view, this will seem a paltry initiative, falling far short of the change that is needed. I agree, which is why I am an ecological pessimist. The changes needed to avert catastrophe are socially and politically infeasible, while the feasible changes we might make are insufficient to avert catastrophe.

This may be a bleak view, but it is not entirely hopeless. In general, it is socially and politically feasible to make the world less bad than it might have been otherwise. Take a simple case. If two people are drowning and I cannot save both, it is better to save one than none. My options are to do nothing or to save one person. Both cases have bad outcomes—one or two deaths—but the less bad outcome is to be preferred. I can be pessimistic here, noting that either course of action has a terrible outcome, but that pessimism does not preclude action, nor does it excuse inaction. It is within my power to dive into the water and save one person. The fact that I cannot save both does not prevent me from saving one. Peter Singer makes a similar point with his famous example of a child drowning in a pond (Singer 1972). It would be neither reasonable nor ethically defensible for a passerby to say, "Well, even if I save this one child, there will be other children who drown, so I might as well let this one die." It is not within the passerby's power to save all drowning children in the world, but that is obviously no reason to decline to save this one child. For Singer, this drowning child is supposed to be analogous to victims of poverty. Just as we ought to help the child, we ought to help those suffering from poverty. A crucial premise in Singer's argument is that, at little cost to ourselves, we *can* prevent a bad thing from happening, whether this is ruining one's shoes by wading into the pond or making a donation to poverty relief. In my experience teaching Singer's argument, students sometimes object to it by pointing out that it is not feasible to prevent *all* poverty. It is true that a single person cannot prevent all cases of suffering due to poverty in the world, but often a single person can prevent some cases of such suffering, so they are obligated to make the attempt. One's inability to eliminate all poverty is no excuse for failing to eliminate some poverty.

The foregoing implies a distinction between what is feasible (e.g., reducing poverty) and what is infeasible (e.g., eliminating all poverty). Benatar provides another example of the latter. He argues, perhaps rightly, that Singer's reasoning entails a certain anti-natalist position (Benatar 2020). Because we have, according to Singer, a duty to prevent bad things from happening, and because procreation causes suffering due to poverty, in many cases, we have an obligation to curtail procreation. That will never happen on a large scale, of course, and I assume Benatar would agree. This is not to find fault with his argument. The reasoning may be impeccable, but our present social conditions do not allow for a global policy that prohibits or greatly restricts procreation. Perhaps it is true that we ought not to reproduce, but that has no bearing on what is reasonably doable in the actual world. Of course, ethicists are free to say that they are not making policy recommendations but only searching for the normative truth, following out logical implications, or something of the sort. That is fair enough, but the gulf between normativity and reality is nonetheless striking, and it provides yet more motivation for pessimism of the moral variety.

This last point mirrors an issue in social-political philosophy, particularly the debate between ideal and non-ideal theorists of justice (see Simmons 2010). Briefly put, ideal justice concerns the principles governing a society in which full compliance with those principles is assumed, whereas non-ideal justice concerns the principles for a society in which full compliance is not assumed. An example is imprisonment. In an ideally just society there would be no justification for putting people in prison, as no one would violate the just laws that are in place. In a non-ideally just society, some people will sometimes break just laws, and imprisonment may then be justified for certain reasons, such as protecting the public or deterring crime. Obviously, we do not live in an ideally just society, so non-ideal theory warrants some attention. Going back to Rawls, some have argued for the priority of ideal theory over non-ideal, suggesting that ideal principles provide guidance for determining the appropriate non-ideal principles (Rawls 1971). Others have argued that we should begin from non-ideal principles, for example, because an ideal theory is ideological (Mills 2005). I will not address that dispute here. The point is only that, insofar as one is doing pure ideal theory, she may ignore the fact of non-compliance in the real world. If she presents a principle of ideal justice, and if it is pointed out that the people will never abide by that principle, she may respond that this is the fault of the people and not of the principle. Of course, whether such ideal theorizing is a valuable activity is a different question.

Once we start paying attention to the issue of non-compliance, we are in the realm of non-ideal theory. Issues of feasibility (social, political, and economic) are of crucial importance for non-ideal theorists. Even supposing that some principle is perfectly just in the ideal sphere, it will be of little use if the application does not work, say because we, vice-ridden people, reject it. I am sympathetic to non-ideal theory, broadly speaking. Whatever may be said for purely ideal theorizing, it is obviously important for us to think about how to organize society in the actual world, perhaps using ideal principles as guiding lights—or perhaps not. One way to think about the matter is as follows. Full compliance with duties of justice is infeasible, both among those with power and the general populace (see human history). Nonetheless, it is feasible to reduce the injustice that would result from this non-compliance, depending on the circumstances. This is a pessimistic view, but not maximally so. We do not live in a just world, but neither do we live in the most unjust world possible. Through certain feasible actions and policies, we can sometimes reduce or at least limit the extent of injustice. Because this is feasible, we have an obligation to make the attempt, just like the passerby who witnesses the drowning child.

In past work, I have made a case like this for research on—and potential deployment of—geoengineering via sulfate aerosol injections (SAI), which would induce global cooling by reflecting a fraction of incoming solar radiation (Morrow and Svoboda 2016; Svoboda 2017). From an ideal-theoretic perspective, the deployment of SAI is almost certainly unjust. For example, SAI threatens to produce unequal distributions of burdens and benefits by causing regional precipitation change, by burdening future generations with the risk of rapid global warming in the wake of sudden cessation of SAI, and by being susceptible to unilateral use (Svoboda et al. 2011). In a global society in which agents fully complied with their duties of justice, we would have pursued emissions mitigation decades ago, and there would be no need for a risky procedure like SAI. In the real world, agents do not fully comply with their duties of justice, so the question is this: Granted that we cannot expect full compliance with justice, might it be the case that SAI deployment is permissible? In other words, although SAI is pretty clearly unjust in an ideal sense, might it be just in a non-ideal sense? Just as imprisoning another human being might be permissible given the fact of violent crime, perhaps deploying SAI can be permissible given the fact that appropriate agents (e.g., high-emitters) have failed to mitigate their emissions.

I concur with the general consensus among climate researchers that we should pursue substantial cuts to emissions, but I also recognize that cuts are occurring only slowly and unreliably. It seems that we are failing to do what justice demands, as we so often do as a species. What then? This is a very important question for an ecological pessimist. Because she has accepted that catastrophe is likely to occur in the future, she cannot help herself to optimism. Indeed, given the morally pessimistic component of ecological pessimism, it would be naive to expect the moral revolution that would be required for us to avert catastrophe. Human history is replete with ghastly injustice, extreme greed, and the pursuit of power, all of which have come at incredible cost to the well-being of parts of humanity and, more recently, non-human nature. It is naive to expect that, any decade now, we will change our ways and become morally serious beings who care about justice in a principled manner. Instead, the pessimist expects us to be as we have always been, easily able to overlook victims outside the relevant tribes, such as future persons, non-human animals, and those living in poverty. A few decent individuals here and there notwithstanding, in general people will help themselves to superficially moral actions when convenient and forego them otherwise. If the pessimist is right about this, then we cannot reasonably expect future climate actions that resemble those demanded by ideal justice.

Despite its own risks of injustice, SAI has the potential to reduce some of the injustices associated with climate change. This is to say that SAI might play a meliorative role, possibly yielding a better (or less bad) state of affairs than other feasible climate policies. For instance, by cooling the planet, SAI could reduce the risks of sea-level rise and subsequent flooding, extreme weather events, reduced agricultural productivity, and drought, among others (Svoboda et al. 2018). These risks all threaten injustice to some persons, especially low-emitters who are especially vulnerable to these aspects of climate change. By reducing these risks, it might be ethically permissible to deploy SAI in certain cases, and some have even argued that doing so might be obligatory (Horton and Keith 2016; cf. Hourdequin 2020). Whatever one thinks about that, in our non-ideal conditions it is at least plausible to consider climate policies involving measures like SAI. We might then compare the merits and demerits of various, feasible policies from the vantage of justice, evaluating them in terms of their respective strengths and weaknesses on that front. We would then have a good ethical reason to prefer the feasible policy that gets us "closest" to actual justice, even if that policy falls short (Svoboda 2016). Considerations of this sort bear a resemblance to just war theory, as some have pointed out (Flord 2023; Fruh and Hedahl 2019). Ordinarily, it is impermissible to shoot other human beings, but under conditions of just war (e.g., defense against aggression), doing so may become permissible. To be clear, I do not claim to know whether SAI will end up comparing favorably to the alternatives, but the ethical case for the potential use of SAI is stronger than many critics suppose.

SAI is just one example of a possible means of meliorating ecological ills. Another, to which I have already alluded, is pursuing feasible mitigation targets. Although I am pessimistic that we can limit warming to 1.5 or 2°C, it is likely feasible to keep warming below (say) four degrees. Two degrees of warming will be bad, but much less bad than four. In saying this is feasible, I am not predicting that we will succeed, but merely observing that holding warming to a moderate level is reasonably doable, given social and political conditions. We may nonetheless fail. Other forms of melioration include limiting the extent of species extinction, ocean acidification, and proliferation of nuclear weapons. Given humanity's moral recalcitrance, no feasible policy will deliver us from these risks, but we might nonetheless reduce them by an appreciable degree. Such efforts are melioristic because they make the world somewhat better (or less bad) than it would have been otherwise.

Why should we work to meliorate these ecological ills, particularly in a way that takes account of our social and political conditions? There are two reasons, broadly speaking, namely self-interest and duty. First, it is simply prudent to reduce risk to ourselves. We all recognize that we have reasons to check roadways for danger, buy various types of insurance, heed warnings of hazardous conditions, and the like. Some ecological risks affect us and things we care about, such as property. Someone with ocean-front real estate has a reason to care about coastal protection. Recreationists have reason to support policies that preserve the practicability such as hikeable trails or streams with the desired species of fish to catch. Arguably, self-interest also extends to persons that we care about. In a sense, it is in one's self-interest to reduce ecological risks to their offspring, perhaps simply because they desire health and happiness for their children, and desire satisfaction is in one's own interest. All of this only goes so far, of course. Self-interest will remain fairly narrow and may give us little reason to care about the health, well-being, rights, or whatever of future generations, currently existing strangers, or non-human animals.

Second, although we often ignore them, we do have moral reasons to reduce ecological risks. Without presuming any particular ethical system (e.g., some form of utilitarianism), it is fairly obvious that the imposition of ecological risk on others is often morally problematic. As usual, I wish to remain ecumenical here, so I will not specify why such risk imposition is morally problematic. Possible reasons include that doing so may be non-consensual, procedurally or distributively unjust, harmful, inimical to the basic functioning of affected persons, or vicious in some way. The only claim I need is that saddling others with risks of catastrophic climate change or nuclear devastation is morally problematic, whatever the reason. If this is true, we have moral reason to reduce such risks, just as we have moral reason to save a drowning child.

The more interesting question is why, at least according to my position, we should take account of relevant social and political conditions when deciding how to act. In keeping with the pragmatic tenor of this chapter, I suggest that we have a moral obligation to consider what is likely to work in a given situation. This is not an appeal to ethical relativism. Even if there are universal moral standards, we nonetheless have an obligation to consider actual circumstances when deciding how best to honor those standards. This should be appealing to anyone who has ever been troubled by Kant's implausible position that we may not lie to a murderer seeking the location of his target (see Korsgaard 1986). It may well be that lying is generally impermissible, but obviously, there are cases in which we ought to lie. Whether that is so will depend on the conditions that hold at the relevant time. This fits with my previous discussion of ideal and non-ideal justice. Under ideal conditions, deployment of SAI might be impermissible. That is far from obvious under certain non-ideal conditions, such as those that might plausibly hold in our actual future. We ought to consider real conditions, because simply relying on universal moral principles risks undermining various moral goods, sometimes the very goods promulgated by the principles in question. For instance, well-meaning activists in the future might oppose SAI on the grounds that it threatens to cause injustice to some low-emitters. Yet the alternatives (e.g., pursuing paltry mitigation and adaptation policies) might involve even greater magnitudes of injustice to low-emitters. When ideally just policies are infeasible in the real world, we need to look to other options in order to limit the moral ills of climate change. In that case, a universal prohibition on SAI might be counterproductive, although this remains uncertain. This is in keeping with the melioristic approach to reducing ecological risk. SAI might be bad, but perhaps less bad than the feasible alternatives.

Peirce's Critique of Philosophy as a Way of Life

There is a potential major problem with this approach, however. In a previous chapter, I defended the idea of philosophy as a way of life, but this may fit poorly with my reliance on pragmatic meliorism. In particular, Peirce himself is very hostile to philosophy as a way of life, as practiced by the ancient Greeks and Romans. Presumably, he would reject my approach for the same reasons. In this section, I consider Peirce's case against philosophy as a way of life. His central charges are that philosophy as a way of life seeks to model practical conduct on the counsels of reason, that reason is currently ill-equipped for that task, and that this leads to dangerous and preposterous consequences. After laying out Peirce's critique, I sketch a mode of philosophy as a way of life that can survive Peirce's charges. Specifically, I suggest that (environmental) philosophy as a way of life should itself be pragmatic, drawing on the lessons of fallibilism and meliorism.

Peirce offers a number of arguments against the rational application of theory to morality, suggesting instead that morality should be grounded in instinct. Peirce maintains that we currently lack the scientific knowledge that would justify a rational structuring of morality. This being the case, philosophically generated moralities cannot be otherwise than dogmatic and dangerous. In this section, I contend that Peirce's critique of what I call "dogmatic-philosophical morality" should be taken very seriously, but I also claim that the purely instinctive morality Peirce endorses is liable to a danger of its own, namely fanaticism. Indeed, Peirce himself recognizes this danger. As an alternative, I sketch a form of "pragmatic morality" that attempts to sidestep the dogmatism of philosophical morality and the fanaticism of instinctive morality. This form of morality avoids philosophical dogmatism by treating extant instincts as the postulates and materials with which they work. It avoids instinctive fanaticism by allowing a role to reason. By exhibiting fallibilism, revisability, pluralism, and meliorism, this type of reasoning can avoid the dogmatism of the philosophical kind of morality Peirce critiques.

To be clear, the terms "philosophical dogmatism" and "instinctive fanaticism" are my own. Although Peirce does not use these terms, they nonetheless capture certain problems that he identifies with different approaches to ethics. Importantly, the point at issue here is not one of metaethics. For instance, although it is tempting to do so, philosophical dogmatism need not be associated with moral cognitivism or the view that moral judgments are beliefs. As Simon Blackburn and Allan Gibbard have shown, non-cognitivists can utilize reasoning in their moral judgments (Blackburn 1993; Gibbard 2008). This opens the possibility that one might be dogmatic in one's normative commitments even if those commitments are best described as desires-like attitudes rather than beliefs. Instead, the point at issue is the *manner* in which we are to conduct ourselves. This is a practical question, not a metaethical one. My claim is that pragmatic morality avoids the problems of both dogmatism and fanaticism, offering a better way of conducting oneself in life.

Peirce distinguishes sharply between practics and morality. Practics is the theoretical science that studies the "conformity of action to an ideal," whereas morality deals with "virtuous conduct, right-living" and cannot "claim a place among the heuretic sciences" (Peirce 1931–58, 1.573). As Juan Pablo Serra points out, "This ideal is neither a socially inculcated one nor a historically or traditionally fixed one" (Serra 2010; see also Massecar 2013). Rather, practics is the normative science that corresponds to action, whereas aesthetics and logic respectively correspond to feeling and thought.[1] There are many questions about what constitutes a Peircean normative science. For instance, James Lizska makes the case that Peirce's stance suggests a kind of normative naturalism, which holds that normative properties are part of nature (Liszka 2014). I am unable to address these various questions here. Instead, I focus on what distinguishes the normative science of practics from morality. Although Peirce does refer to morality as a science (ibid.), it is not one that makes original discoveries, i.e., it is not "heuretic." In fact, it is unwise to mix practics and morality. The latter is "the folklore of right conduct," the "traditional wisdom of ages of experience" that becomes ingrained in one's conscience through upbringing. It is dangerous to even reason about morality, "except in a purely speculative way." "Hence, morality is essentially conservative" (Peirce 1931–58, 1.50). This "purely speculative" manner of reasoning about morality is presumably the science of practics. For Peirce then, moral conduct and normative science are distinct, quite separate pursuits, neither of which influences the other. As Vincent Potter notes in reference to Peirce,

To say that knowledge of normative science would directly and in itself either help one to think more correctly or to live more decently or to create more artistically, would be like saying that a knowledge of mechanics involved in a game of billiards would allow us to become a master player (cf. e.g. Peirce 1931–58, 2.3).

(Potter 1967: 26)

Although Peirce cautions that applying the results of practics to morality is extremely perilous, he is not opposed to it in principle: "I do not say that philosophical science should not ultimately influence religion and morality; I only say that it should be allowed to do so only with secular slowness and the most conservative caution" (Peirce 1931–58, 1.620). Yet although Peirce does not in principle forbid the application of practics to morality, doing so is usually a fallacious and dangerous exercise. Peirce notes two peculiar characteristics of any professional thief: "first, an even more immense conceit in his own reasoning powers than is common, and second, a disposition to reason about the basis of morals." Evidently, Peirce thinks the thief is guilty of self-conceit, special pleading, and cynicism in misapplying pseudo practics to morality. "Hence, ethics, which is reasoning out an explanation of morality is… composed of the very substance of immorality" (Peirce 1931–58, 1.666).[2] I use the term "dogmatic-philosophical morality" to denote any system of morals generated by a dogmatic application of science to conduct.

In his lecture, "Philosophy and the Conduct of Life," Peirce proclaims himself "an Aristotelian and a scientific man, condemning with the whole strength of conviction the Hellenic tendency to mingle Philosophy and Practice" (Peirce 1931–58, 1.618). In this work, Peirce forcibly argues that morality should be left to instinct and that reasoning about morality can and often does lead to dangerous consequences. Before considering this lecture, however, the context of the work must be addressed. As a part of his Cambridge Lectures of 1898, Peirce delivered this talk at Harvard at the invitation of William James, who had urged Peirce to lecture on matters of "vital importance" rather than on less interesting topics, such as "formal logic." James went so far as to write Peirce the following: "Now be a good boy and think a more popular plan out" (Peirce 1998, 505*n*15). Understandably, Peirce seems to have taken exception to this, hence the ironical use of "matters of vital importance" throughout much of the lecture. How should this context affect one's reading of the lecture? Cheryl Misak claims that the Cambridge Lectures "are not the best place for discerning Peirce's considered view about science and vital matters," because Peirce's anger at James causes him to overstate his case (Misak 2004, 163). To some extent, this assessment seems appropriate. For one thing, Peirce's "considered view" (after 1903, at least) includes the normative science of ethics as part of philosophy, something he disavows in "Philosophy and the Conduct of Life." However, this lecture is not devoid of interesting, plausible arguments and claims of its own. Moreover, Peirce does not seem to substantially change the view expressed in this lecture that I explore most closely, namely that morality should be grounded in instinct rather than in reason. In fact, this claim is consistent with much that Peirce writes elsewhere, for example:

Invariably follow the dictates of Instinct in preference to those of Reason when such conduct will answer your purpose: that is the prescription of Reason herself. Do not harbor any expectation that the study of logic can improve your judgment in matters of business, family, or other departments of ordinary life. Clear as it seems to me that certain *dicta* of my conscience are unreasonable, and though I know it may very well be wrong, yet I trust to its authority emphatically rather than to any rationalistic morality. This is the only rational course.

(Peirce 1931–58: 2.177)

Peirce adds to this, "The best plan, then, on the whole, is to base our conduct as much as possible on instinct, but when we do reason to reason with severely scientific logic" (Peirce 1931–58: 2.178). Hence, although Peirce's *rhetoric* in "Philosophy and the Conduct of Life" may be ironically or even sarcastically tinged on account of James' patronizing remarks, and although he later makes room in philosophy for ethics as a *theoretical* science, Peirce nonetheless seems committed to the closely related positions that morality should be grounded in instinct and that reason should not meddle in morality, because these claims are not isolated to this lecture alone. In fact, as Richard Atkins argues, the lecture can be read as a critique of James' assumption that philosophy should concern itself with "matters of vital importance," and this gives us all the more reason to take its arguments seriously (Atkins 2016).

In "Philosophy and the Conduct of Life," Peirce outlines five traits exhibited by philosophy, which distinguish it both from practice and the special sciences: (1) unlike mathematics, it searches for real truth, (2) it takes its premises from experience, (3) it investigates the reality of potential beings, not just existing beings, (4) its premises are not specially observed facts but rather the universal phenomena that "saturate" all experience, and (5) its metaphysical conclusions show how things must be.[3] Peirce divides philosophy into metaphysics and logic. Metaphysics studies "being in general, its laws and types." Logic studies "thought in general, its general laws and kinds" (Peirce 1889, 36). Though he would later change his mind to a significant degree,[4] at this point, Peirce excludes ethics from philosophy for two reasons. First, although it is "the science of the end and aim of life," ethics deals only with "a special department of experience," namely the "psychical." For this reason, it does not display trait (4) of philosophy, because it does not deal with universal phenomena. Second, ethics seems to be an art, or at least one of the "theories of the arts." Hence, it is a "concrete" theoretical science, whereas philosophy is "the most abstract of all the real sciences" (Peirce 1931–58, 36).

If one accepts this characterization of philosophy, then one can agree with Peirce that philosophy is in an "infantile condition" (Peirce 1931–58, 1.620). In order to be properly philosophical, any knowledge claim would need to satisfy the following necessary conditions, corresponding to traits (1), (2), (4), and (5) above: it must (1) concern real truth, (2) follow from premises drawn from experiences that are (4) universal phenomena, and (5) hold with metaphysical necessity. I do not include trait (3) as a *necessary* condition for any claim to be properly philosophical, because Peirce says that philosophy investigates potential beings *in addition* to existing ones, which means that a particular claim need not deal with potential beings in order to be philosophical. To claim knowledge about metaphysically necessary, universal phenomena is a serious business. To suggest a system of morals based on claims about metaphysically necessary, universal phenomena is an even more serious business. Such an application of philosophy to practice, if not strongly justified, would be premature at best, and catastrophic at worst. The result of such an inappropriate application of philosophy to practice is what I am calling "dogmatic-philosophical morality," i.e., a system of conduct that is formulated according to unjustified philosophical claims. Not only does the Peirce of 1898 exclude ethics from philosophy on the grounds that it does not concern universal phenomena, but he also suggests that philosophically formulated moralities are likely to have very dangerous practical consequences. Although the Peirce of 1903 and later makes room for ethics in philosophy, he seems to maintain his stance that applying philosophy to morality is a very hazardous undertaking.

In keeping with his own advice, Peirce does not attempt to closely delineate the source and extent of morality. The "best opinion" is that morality "has its root in the nature of the human soul, whether as a decree of reason, or what constitutes man's happiness, or in some other department of human nature" (Peirce 1931–58, 2.156). Peirce here leaves the exact nature of morality, its relation to reason, and its relation to happiness as open questions. The view that morality is rooted in human nature is only an opinion, a more or less plausible conjecture. It is interesting that Peirce here considers morality as grounded in the "human soul," whereas elsewhere he treats morality as a matter of up-bringing: "[a] man is brought up to think he ought to behave in certain ways," (Peirce 1931–58, 1.50). Of course, there need be no conflict between the morality of upbringing and the morality rooted in the soul, since one might be brought up to conform to the morality suggested (perhaps implicitly) by the latter. Peirce seems to treat morality as encompassing both aspects. "There is probably no special instinct—using this word in a sense in which it shall embrace traditional as well as inherited habits—for rationality, such as there is for morality" (Peirce 1931–58, 2.160). This passage is important for two reasons. First, Peirce suggests that there is an instinct for morality. Second, this instinct encompasses both "traditional" and "inherited" habits. The import of "traditional" and "inherited" is not immediately clear. However, as noted above, Peirce characterizes morality as the *traditional* wisdom of ages of experience" (Peirce 1931–58, 1.50). This would suggest that the traditional habits of moral instinct mentioned in 2.160 are the habits rooted in the "ages of experience" of one's community mentioned in Peirce 1931–58, 1.50. Hence, traditional habits seem to be those whose origin is the custom of one's community. "Inherited" habits, on the other hand, seem to be those that are *acquired* irrespective of the custom of one's community. Because it is difficult to differentiate an inherited habit from one "due to infantile training and tradition," Peirce uses the word "instinct" to cover both types of habit (Peirce 1931–58, 2.170). In short, moral instinct includes both habits ingrained by the customs of one's community and habits acquired by other means. Accordingly, they, to some extent, must admit of variability, change, improvement, and degeneration.

A Pragmatic Approach to Philosophy as a Way of Life

There is no question that Peirce is highly suspicious of philosophical meddling in morals. Nonetheless, he also notes that "morality, doctrinaire conservatist that it is, destroys its own vitality by resisting change, and positively insisting, This is eternally right: That is eternally wrong." Peirce adds, "Like any other field, more than any other, it [morality] needs improvement, advance. Moral ideas must be a rising tide, or with the ebb foulness will be cast up." He concludes,

> The practical side of ethics is its most obviously important side; and in practical matters, the first maxim is that everything may be exaggerated. [...] The moral spirit may very easily be carried to excess: all the more so, that the essence of that spirit is to insist upon its own absolute autocracy.
>
> (Peirce 1931–58, 2.198)

While it is dangerous to apply prematurely the results of practics to morality, moral fanaticism presents its own serious hazards. But if the work of practics is not sufficiently advanced to offer much help to morality, and if humans have nothing except moral instinct to ground and guide their conduct, how can such moral fanaticism be avoided? In the remainder of this section, I sketch a possible solution to this problem that takes seriously Peirce's criticisms of philosophical morality. Relinquishing claims to certainty and universality, one can negotiate both the dogmatism of dogmatic-philosophical morality and the fanaticism of unhindered moral conviction. Invoking pragmatic traits like fallibilism, pluralism, and revisability, one can deploy reason to think intelligently about the traditional and inherited habits that comprise one's moral instincts. I attempt to show that reasoning about morality need not be dogmatic but can suggest richer modes of living than either dogmatic-philosophical morality or purely instinctive morality.[5]

As mentioned above, I use the term "dogmatic-philosophical morality" to denote any dogmatic application of philosophy, including practics, to conduct. I shall use the term "pragmatic morality" to denote the non-dogmatic reasoning about morals I now wish to sketch. In such a pragmatic morality, reason would work *with* instincts. Peirce seems amenable to this view. In the 1898 lecture, "Philosophy and the Conduct of Life," Peirce says that in "vital crisis" reason appeals to instinct (Peirce 1931–58, 1.630). In theoretical science, appeals to instinct are inappropriate, except insofar as they suggest procedures that might be tried and tested. In "human affairs," conversely, reason itself suggests the "supremacy of sentiment,"[6] just as it refuses anything other than a suggestive role for sentiment in theoretical matters (Peirce 1931–58, 1.634). Interestingly, Peirce notes that instinct is amenable to "development and growth" no less than reasoning, though the former should be a slow and careful development when the instinct is 'vital.'" Moreover, this development of instinct "chiefly takes place through the instrumentality of cognition." In this manner, the "eternal forms" known through the sciences gradually come to influence the lives of human beings, and this "not because they involve truths of merely vital importance, but because they are ideal and eternal verities" (Peirce 1931–58, 1.648).[7]

There are at least two plausible interpretations of this passage in Peirce 1931–58, 1.648, neither of which needs to completely exclude the other. On the one hand, Peirce may be reinforcing the claim I discuss above, namely that practics should be applied to morality only after the former has made significantly more progress than it currently has, after which it still must be applied with acute caution. On the other hand, Peirce may be suggesting that cognition or reason[8] can play a helpful role vis-à-vis morality—perhaps by softening the fanaticism of pure moral instinctiveness and by bringing various instincts into greater harmony—even before practics has made significant progress. Indeed, some readers have taken Peirce to be committed to moral cognitivism, which holds that moral judgments are mental states that are true or false, namely beliefs.[9] Whichever of these two interpretations of Peirce 1931–58, 1.648 one adopts, it must account for the following four claims made in the passage: (1) vital instinct admits of growth and development, (2) the development of vital instinct is slow, (3) cognition or reason is the primary instrument that effects the growth and development of vital instinct, (4) the truths of science will eventually influence morality[10] because of their eternal verity and not for any other reason. I consider each of these two interpretations in turn.

The first interpretation of Peirce 1931–58, 1.648, which reads it as forbidding the application of reasoning to vital instincts until science (especially practics) has made sufficient progress, accounts for Peirce's four claims in the following way. The numbers in parentheses correspond to claims (1)–(4) directly above. While (1) vital instincts are indeed amenable to development, and while (3) reasoning is the instrument that effects this development, (4) such reasoning should influence morality only in virtue of scientific truths not yet known, hence (2) the slow and gradual development of vital instincts. In this interpretation, reasoning about vital instinct is appropriate only when science is sufficiently advanced to justify such reasoning. Absent this advanced state of science, one should never seek to alter morality by reasoning about it. Even when science is sufficiently advanced, reasoning should alter the vital instincts that comprise morality in only a slow and gradual manner.

The second interpretation, which reads the passage as endorsing a kind of reasoning about instincts that does not require that reasoning to be based in a highly progressed science, accounts for claims (1)–(4) in a different manner. Since (1) vital instincts admit of growth and development independently of (4) the currently unknown truths of science that would justify a *deep-reaching* overhaul of those vital instincts, (3) a provisional and cautious sort of reasoning can be used as an instrument to achieve (2) a slow and gradual development of the vital instincts that comprise morality. In this interpretation, only the "eternal verities" of science can justify a *significant* and *deep-reaching* alteration of the morality suggested by vital instinct. Nonetheless, lacking knowledge of such verities, one can meanwhile reason cautiously about vital instincts, and this reasoning can suggest slow, gradual ways of changing those instincts for the improvement of morality. Such change should not be deep-reaching, as only scientific knowledge not currently had could justify such significant alteration to morality. However, unlike the first interpretation, this one grants a role to reason vis-à-vis vital instinct, despite the fact that science has not progressed enough to justify more than provisional reasoning. I call the result of this cautious working of reason upon vital instincts "pragmatic morality."

As a matter of exegesis, I grant that the first interpretation better coheres with those passages in which Peirce maintains a deep division between the sciences (which include philosophy) and morality (e.g., Peirce 1931–58, 1.50, 1.666). The first interpretation also provides a view of the relation between science and morality that offers an antidote to moral fanaticism, albeit a long-delayed one. As discussed above, Peirce claims that morality "needs improvement, advance," lest it degenerate into a trenchant dogmatism of its own (Peirce 1931–58, 2.198). In keeping with the first interpretation of Peirce 1931–58, 1.648, Peirce might argue that the "eternal verities" of science eventually soften and reform the conservative dogmatism of morality. Once science has made sufficient progress in some area, the theoretical knowledge it yields might be applied to morality so as to improve it, not least by curbing the zeal with which it clings to certain instincts. While this does seem to be Peirce's position, his view has the weakness that there is nothing to be done about any aspect of moral fanaticism *prior* to science's making sufficient progress in some relevant area.

He maintains that science should influence morality "only with secular slowness and the most conservative caution" (Peirce 1931–58, 1.620). Following the first interpretation of Peirce 1931–58, 1.648, this would require that one obeys one's vital instincts in all matters of morality upon which science is not advanced enough to pronounce. But since this instinctive morality "destroys its own vitality by resisting change," there seems to be no protection against moral fanaticism (Peirce 1931–58, 2.198). One might counter that the risk of fanaticism in instinctive morality is less dangerous than dogmatic-philosophical morality. The latter often imposes new injunctions on practical activity, which, if followed, can lead to disastrous, unforeseen consequences. Moreover, there is no guarantee that any dogmatic-philosophical morality is even practicable, since such morality cannot be tested prior to its endorsement. Conversely, instinctive morality has survived the crucible of human history, so it must be practicable and minimally satisfactory. On these grounds, one might hold that instinctive morality, though prone to fanaticism, is nonetheless preferable to dogmatic-philosophical morality. I grant that this is true. Nonetheless, in the absence of properly applicable scientific knowledge, one wishes there were some third option between instinctive morality and dogmatic-philosophical morality.

The second interpretation of 1.648, though less plausibly attributable to Peirce, offers just such a middle ground between the fanaticism of instinctive morality and the dogmatism of philosophical morality. In this interpretation, one can reason about vital instincts in a cautious, responsible, and tentative manner without needing to know the truths of science. This kind of reasoning about morality relinquishes the claim to the certainty that dogmatic-philosophical morality always asserts, hence avoiding dogmatism and making the resulting pragmatic morality much less dangerous. It also avoids moral fanaticism by intelligently dealing with instincts and questioning the zeal with which certain instincts are espoused. I call this type of reasoning about moral instincts "pragmatic" because it displays a number of qualities associated with pragmatism, such as fallibilism and pluralism. This pragmatic reasoning can (1) consider the likely consequences of abiding by a particular instinct or set of instincts, (2) compare and weigh instincts that suggest conflicting paths of conduct, and (3) propose certain orderings of the relative importance of various instincts.[11] Whereas dogmatic-philosophical morality results from unjustified claims to certainty, and instinctive morality offers no defense against fanaticism, pragmatic morality results from taking pre-existing vital instincts as material to be thought through, intelligently ordered, and adhered to cautiously.[12] It is clear that this second interpretation of 1.648 conflicts with the first one. The first interpretation is more consistent with the rest of Peirce's claims in "Philosophy and the Conduct of Life," but I wish to examine the pragmatic morality suggested by the second one. I argue that, although he does not endorse it, Peirce could approve a pragmatic morality, because it avoids the problems of dogmatic-philosophical morality and offers correctives for some of the dangers of purely instinctive morality.

Peirce's chief objection to the mingling of philosophy and the other sciences with morality seems to be that the dogmatism of the former produces unacceptable *practical* effects for morality. Indeed, Peirce says that reason itself recommends that one trust in vital instinct when it comes to moral matters, thereby eliminating itself from being put to moral use (Peirce 1931–58, 1.634). Peirce criticizes the "early Greek philosopher, such as we read about in Diogenes Laertius," whose "conduct should be in marked contrast with the dictates of ordinary common sense" (Peirce 1931–58, 1.616). In this view, the Greek philosophers prematurely used theories about the cosmos and human beings to justify moralities in deep contrast with those suggested by vital instinct. On this point, Richard Shusterman notes that philosophy as theory and philosophy as an art of living were complementary for Greek philosophers (Shusterman 1997, 4).[13] He adds that it would be difficult to separate "Epicurean natural theory" about atoms and the void from the Epicurean "art of living," or Stoic ethics from its "philosophical theory ... that viewed the whole natural world as a perfect, living organic unity" (Shusterman 1997, 4).[14] The Epicurean insistence that one should fear neither the gods nor death seems based on the atomistic worldview, whose mechanisms leave no room for divine interference and which make death merely the complete dissolution of the human being. Hence, the gods cannot afflict human beings, and there is no state of suffering after death. Convinced of these truths on the basis of metaphysical theory (science), one can confidently pursue an art of living (morality) that intelligently cultivates pleasure. Similarly, the Stoic insistence that one should accept the whole of nature and live in accordance with it seems based on the worldview that nature is a perfect, providential whole. The art of living whereby one accepts nature and accords with it is justified by the theory that ensures nature is something worthy of being accepted and accorded with.

Peirce is suspicious of this very relationship between theory and arts of living, between science and morality. If there were very good reason to believe that either the Epicurean or Stoic theory was true, then it might be appropriate to adopt its corresponding art of living. But since Peirce does not think there is sufficiently good reason to hold either theory true, the pursuit of either art of living cannot be otherwise than dangerous and ridiculous, producing "one of the most amusing curiosities of the whole human menagerie," the philosopher who feels compelled to conduct himself in direct opposition to common sense (Peirce 1931–58, 1.616). This leads to a dogmatic-philosophical morality. Reason is employed to construct flimsy doctrines about the universe and human beings, upon which a dogmatic morality is constructed. Alternatively, I suggest that reason should abstain from making unjustified claims about the cosmos and morality, but that reason can still play a role in moral matters. Such pragmatic reasoning need not be dogmatic, because it need not assume universal truths about the cosmos and human beings. Pierre Hadot suggests the following:

> The same spiritual exercises can, in fact, be justified by extremely diverse philosophical discourses. These latter are nothing but clumsy attempts, coming after the fact, to describe and justify inner experiences whose existential density is not, in the last analysis, susceptible of any attempt at theorization or systematization.
>
> (Hadot 1995, 21)[15]

Hadot reverses the order of influence. According to Peirce's account, the ancient philosopher first claims to know universal truths about the cosmos and human beings, after which she constructs a morality justified by these supposed truths. For Hadot, it seems one adopts a morality (a set of "spiritual exercises") first, after which one seeks universal truths about the cosmos and human beings ("philosophical discourses") to justify or otherwise validate this morality. I leave aside the historical question of whether Hadot's view adequately describes what Greek philosophers actually did or what they understood themselves to be doing. Instead, I take Hadot's claim as a clue that can help sketch a pragmatic morality that avoids the dogmatism of philosophical morality and the fanaticism of purely instinctive morality.

Hadot's position avoids philosophical dogmatism by minimizing the role of philosophical discourse, but it immediately faces two serious objections. If philosophical discourse (Shusterman's theory, Peirce's science) does not inform one's choice in adopting a set of spiritual exercises (Shusterman's art of living, Peirce's morality), then (1) how does one acquire a set of such exercises, and (2) what keeps this acquisition from being arbitrary? Peirce agrees that philosophical discourse or science *should not* determine one's morality, suggesting that instead vital instinct should serve this function. This constitutes a response to objection (1)—one acquires one's morality (a set of spiritual exercises) from one's vital instincts. Yet it appears that Peirce's instinctive morality may not adequately answer objection (2), because one's morality is constituted by the vital instincts one only *happens* to have. Hence, instinctive morality is arbitrarily acquired in that it results from contingent instincts rather than from voluntary or rational procedures. A defender of Peirce might make the following two rejoinders. First, although arbitrary in the sense that the individual does not volitionally or rationally validate it, instinctive morality is the "traditional wisdom of ages of experience" grounded in one's community and its history, which means it must be at least minimally serviceable (Peirce 1931–58, 1.50). Second, granted that instinctive morality is arbitrary in the above sense, it remains the best kind of morality one can currently hope for, because science (in particular, practics) is not sufficiently progressed to let one impose rational or volitional procedures on morality in a justified and responsible manner. One might cede both these rejoinders and be resigned to the fact that there is no better option than instinctive morality, which is at least minimally satisfactory. But it would no doubt be better if the individual could reason about moral matters while being neither philosophically dogmatic nor arbitrarily committed to the morality one happens to find as one's own. Shusterman offers a clue in this direction:

> For even if we doubt that every art of living entails a full-blown philosophical theory and every theory expresses a way of life, we surely should build our art of living on our knowledge and vision of the world, and reciprocally seek the knowledge that serves our art of living.
>
> (Shusterman 1931–58, 4)

Morality need not be justified by a "full-blown philosophical theory" in order to be informed by the rationality of individuals in a community. Moreover, individuals may reason about their morality in a manner that might require the emendation or even complete abandonment of that morality. Hence, there can be a kind of morality that avoids both philosophical dogmatism and instinctive intractability. I now sketch some of the features of this pragmatic morality.

First and most importantly, this pragmatic reasoning and the pragmatic morality resulting from it are both fallibilist. Rather than seeking universal moral truths that are justified by other universal truths about the world and human beings, pragmatic reasoners consider the relative plausibility of various moral claims. Hence, pragmatic morality deals with probability rather than certainty. Taking Peirce's vital instincts as so many postulates about morality, the pragmatic reasoner considers matters[16] such as the following: (1) the mutual coherence of any set of vital instincts, (2) the likely consequences of adhering to various combinations of vital instincts, (3) information surrendered by past and present experiments of acting on various of those instincts,[17] (4) possible alternatives to current instincts, (5) the likely consequences of those alternatives, and (6) whether the risk of altering a morality is worth taking. In all these functions, the reasoner can remain sincerely fallibilist, shirking inappropriate certitude and deferring to vital instinct when her pragmatic reason seems unable to proceed responsibly. This deployment of reason is not scientific or theoretical in the sense of seeking certainty but is rather closer to *phronesis*. The pragmatic morality that results from this phronetic, pragmatic reason avoids the dogmatism of philosophical morality by declining to invest undue certitude in any moral claim. Pragmatic morality avoids the fanaticism of purely instinctive morality by reasoning about instincts and thereby curbing the zeal with which they assert themselves. The resulting morality is well-grounded in time-tested instincts, but it also admits of improvement.

This leads to the second important trait of pragmatic morality, its revisability, which follows from its fallibilism. Given that views about morality are not treated as certain but only as more or less probable, the pragmatic reasoner is always open to revising any position, perhaps drastically. This revision may consist in granting more influence to vital instinct, or it may consist in adopting newer moral habits at the suggestion of pragmatic reason. Either way, the pragmatic reasoner remains committed to honestly evaluating what information is available and proceeding accordingly. The proper procedure may require her to put reason away and abide by instinct, but it might also require her to revise her morality by altering those instincts. It is unlikely that the pragmatically rational course will call for a deep-reaching overhaul of one's morality, but one must be prepared for such revision should pragmatic reason suggest it. If one denies that one's morality is revisable, then one risks either the fanaticism of instincts or the dogmatism of philosophical morality, provided that one is not scientifically or theoretically justified in that philosophical morality.

Third, pragmatic morality admits of pluralism, and this again in virtue of its fallibilism. Since no moral position is held with certainty, the pragmatic reasoner can view a multiplicity of moralities as appropriate. This does not commit the pragmatic reasoner to moral relativism, because she need not hold that inconsistent moral claims are all true. Instead, she may claim (1) that human inquirers currently lack the scientific or theoretical knowledge that would permit a final adjudication of inconsistent moral claims, and (2) that in our current condition plausible cases can be made for inconsistent moral claims. In any pragmatic morality, each moral position is maintained as more or less probable vis-à-vis other positions, but such a position is inherently revisable. This being the case, the pragmatic reasoner does not feel justified in refusing every other, inconsistent morality.

Fourth, pragmatic morality is centrally committed to meliorism or the position that human effort can improve human life. This is not to deny that the purely instinctive morality recommended by Peirce might also be committed to meliorism. However, purely instinctive morality seems liable to stagnation and intractability. If one is morally committed only to what vital instinct suggests, then it is hard to see how a morality could ever progress, unless sufficient progress should be made in science. This morality's intractability would seem to hamper the effectiveness of its melioration of human life, since it can only pursue the improvement thereof within the framework that its instincts happen to offer. On the other hand, being fallibilist and revisable, pragmatic morality admits of self-improvement and intelligent direction. Although it takes vital instincts as its starting points, pragmatic morality's revisability lets it be developed in a manner instinctive morality does not allow. The pragmatic reasoner can tweak vital instincts to bring them into greater coherence, produce more favorable consequences, abandon those that have been found harmful, and suggest new ones that beneficially supplement the rest. Being thereby more flexible than instinctive morality, pragmatic morality can also better adapt itself to changing conditions and new situations. For all these reasons, the meliorism of pragmatic morality seems much more effective and far-reaching than any meliorism that might be found in instinctive morality, at least until the science of practics is insufficiently advanced to responsibly influence morality. Until such time, it seems the best hope for the melioration of human life lies in pragmatic morality.

The purpose of this section has been to offer a preliminary sketch of a pragmatic morality that takes seriously Peirce's criticisms of both the dogmatism of philosophical morality and the fanaticism of instinctive morality. Peirce seems to favor instinctive morality, despite the penchant for fanaticism Peirce himself recognizes in it. This is a respectable position, given that Peirce does not believe that science is advanced enough to permit a rational overhaul of morality. However, I have attempted to draw a rough outline of a pragmatic morality that sidesteps both dogmatism and fanaticism. By exhibiting the traits of fallibilism, revisability, pluralism, and meliorism, this pragmatic morality permits reasoning about morals in a responsible, non-dogmatic fashion. Moreover, by allowing a role for vital instincts as the material or postulates of morality, pragmatic morality does not endorse potentially dangerous, deep-reaching renovations of morals. If successful, this outline has suggested a kind of morality in which vital instinct and reason form a complementary relationship.

All of this allows us to draw an important lesson for the ecological pessimist. Meliorism is not just a matter of public policy. It also pertains to one's own life. Just as we cannot reasonably hope to achieve a utopian ideal in the social-political realm, we cannot reasonably hope to achieve perfectly good lives, but in neither case is this cause for quietism. We can improve the world and ourselves to some degree, and those are tasks worth pursuing.

Notes

1 Peirce's views about the three normative sciences (aesthetics, practics, and logic) are quite interesting. Peirce makes aesthetics the first normative sci-

ence, followed by practics, then logic. However, aesthetics and logic are not directly relevant to this chapter, hence I do not consider them closely. See Potter (1967).

2 Here "ethics" is functionally synonymous with "practics" as it appears elsewhere.

3 This portion of the lecture is not included in the *Collected Papers*. See Peirce (1889, 35).

4 Peirce revised his position by 1903, after which he viewed philosophy as comprised of phenomenology, normative science, and metaphysics, normative science itself being comprised by aesthetics, ethics, and logic. See Peirce (1889, 506*n*28).

5 I do not mean to suggest that Peirce would be opposed to such an approach. On the contrary, several passages suggest he might approve of it. Nonetheless, Peirce does not offer much in the way of achieving a pragmatic morality, hence my attempt to sketch some of the traits such a morality would display.

6 Throughout my reading of "Philosophy and the Conduct of Life," I take "instinct" and "sentiment" to be functionally synonymous.

7 This is perhaps the most overt shot Peirce takes at James in this lecture.

8 Peirce says "cognition" in this passage, but I see no textual evidence that he means by it anything significantly different from what he means by "reason" elsewhere in the same lecture. On the contrary, immediately prior to his claim that the development of instinct or sentiment "takes place through the instrumentality of cognition," Peirce says that the development of instinct or sentiment "takes place upon lines which are altogether parallel to those of reasoning," suggesting that his use of "cognition" and "reason" or "reasoning" are indeed synonymous.

9 For further discussion, see Massecar (2014).

10 Peirce says only that scientific truth will "influence our lives," but the context makes it clear that he means that scientific truth will gradually come to inform the practical activity of human beings in vital matters, what he elsewhere calls "morality."

11 Of course, one could list many other tasks of which pragmatic reasoning about instincts is capable, but these three are especially relevant to my task of sketching a pragmatic morality that avoids the problems of both dogmatic-philosophical and purely instinctive moralities.

12 To be viable, the pragmatic kind of morality must answer a number of challenges, such as the two following. Can communities and/or individuals function with fallibilist attitudes toward their moralities, or do they need certitude? Absent scientific knowledge, on what basis does one reason about instincts? I offer some responses to these questions below.

13 In good pragmatic fashion, Shusterman treats philosophical theory as a kind of practice, one concerning "the formulation or criticism of general, systematic views about the world." The philosophical "art of living," conversely, is the practice whereby one lives well.

14 Shusterman uses terms like "art of living," "life practice," and (as here) "ethics" to denote what Peirce means by "morality."

15 Hadot's "spiritual exercises" and "philosophical discourses" are respectively equivalent to Peirce's "morality" and "science," as well as Shusterman's "art of living" and "theory."

16 Again, this list need not exhaust all the functions of pragmatic reasoning about morality. I only address those that are immediately salient for dealing with Peirce's criticisms. One of the strengths of pragmatic reasoning about

morality is that it itself admits of development, growth, correction, improvement, etc., so trying to list all its functions *a priori* would be a misguided endeavor.

17 While the pragmatic reasoner would not endorse radical experiments in morality that might be quite dangerous, it is undeniable that such experiments (intentionally undertaken or not) are made by some people, ranging from heavy drug use and unusual sexual activity to monastic practices of fasting and meditation. Even if a pragmatic morality would not endorse such experiments, the pragmatic reasoner should still study them and learn what she can. This increased knowledge cannot hurt pragmatic morality, and it may improve it, either by reinforcing certain vital instincts or by suggesting that those instincts might be safely modified after all.

References

Atkins, Richard Kenneth. *Peirce and the Conduct of Life: Sentiment and Instinct in Ethics and Religion*. Cambridge: Cambridge University Press, 2016.

Beiser, Frederick C. *Weltschmerz: pessimism in German philosophy, 1860–1900*. Oxford: Oxford University Press, 2016.

Benatar, David. "Famine, Affluence, and Procreation: Peter Singer and Anti-Natalism Lite." *Ethical Theory and Moral Practice* 23 (2020): 415–431.

Simon Blackburn. "Attitudes and Contents." In *Essays in Quasi-Realism*. New York: Oxford University Press, 1993, 182–197.

Dewey, John. *John Dewey: The Middle Works, 1899–1924, Volume 12*. Edited by Jo Ann Boydson. Carbondale: Southern Illinois University Press, 2008.

Floyd, Rita. "Solar Geoengineering: The View from Just War/Securitization Theories." *Journal of Global Security Studies* 8, no. 2 (2023): ogad012.

Fruh, Kyle, and Marcus Hedahl. "Climate Change Is Unjust War: Geoengineering and the Rising Tides of War." *The Southern Journal of Philosophy* 57, no. 3 (2019): 378–401.

Gibbard, Allan. *Thinking How to Live*. Cambridge: Harvard University Press, 2008.

Pierre, Hadot. *Philosophy as a Way of Life*. Translated by Michael Chase. Malden. Blackwell, 1995.

Horton, Joshua, and David Keith. "Solar Geoengineering and Obligations to the Global Poor." In *Climate Justice and Geoengineering: Ethics and Policy in the Atmospheric Anthropocene*, edited by C. J. Preston. Cambridge, MA: NBER, 2016, 79–92.

Hourdequin, Marion. "Climate Change, Climate Engineering, and the 'Global Poor': What Does Justice Require?." In *The Ethics of "Geoengineering" the Global Climate*. London: Routledge, 2020, 41–59.

James, William. *Pragmatism*. New York: Longmans, Green and Co, 1907.

Kim, Jaegwon. "The American Origins of Philosophical Naturalism." *Journal of Philosophical Research* 28, no. Supplement (2003): 83–98.

Korsgaard, Christine M. "The Right to Lie: Kant on Dealing with Evil." *Philosophy & Public Affairs* 15 (1986): 325–349.

Liszka, James. "Pragmatism and the Ethic of Meliorism." *European Journal of Pragmatism and American Philosophy* 13 (2021): XIII-2.

Liszka, James. "Peirce's Idea of Ethics as a Normative Science. Charles S. Peirce Society Presidential Address. *Transactions of the Charles S. Peirce Society* 50, no. 4 (2014): 459–479.

Margolis, Joseph. "Pragmatism without Foundations." *American Philosophical Quarterly* 21, no. 1 (1984): 69–80.

Massecar, Aaron. "The Fitness of an Ideal: A Peircean Ethics." *Contemporary Pragmatism* 10 (2013): 97–119.

Massecar, Aaron. "Peirce, Moral Cognitivism, and the Development of Character." *Transactions of the Charles S. Peirce Society: A Quarterly Journal in American Philosophy* 50, no. 1 (2014): 139–161.

Mills, Charles W. "'Ideal Theory' as Ideology." *Hypatia* 20, no. 3 (2005): 165–183.

Misak, Cheryl. "C. S. Peirce on Vital Matters." In *The Cambridge Companion to Peirce*, edited by Cheryl Misak. Cambridge: Cambridge University Press, 2004, 150–174.

Morrow, David, and Toby Svoboda. "Geoengineering and Non-Ideal Theory." *Public Affairs Quarterly* 30, no. 1 (2016): 83–102.

Nolt, John. "How Harmful Are the Average American's Greenhouse Gas Emissions?." *Ethics, Policy and Environment* 14, no. 1 (2011): 3–10.

Nolt, John. "Casualties as a Moral Measure of Climate Change." *Climatic Change* 130 (2015): 347–358.

Peirce, Charles Sanders. "Meliorism." In *Century Dictionary, Volume 5*, edited by William Whitney. Century Co, 1889–1991.

Peirce, Charles Sanders. *The Collected Papers of Charles Sanders Peirce*, 8 vols. Edited by Arthur Burks, Charles Hartshorne, and Paul Weiss. Cambridge: Harvard University Press, 1931–1958.

Peirce, Charles Sanders. *The Essential Peirce*, vol. 2. Edited by Nathan Houser, et al. Bloomington and Indianapolis: Indiana UP, 1931–1958.

Potter, Vincent. *Charles S. Peirce on Norms and Ideals*. Worcester: University of Massachusetts Press, 1967.

Rawls, John. *A Theory of Justice*. Cambridge: Harvard University Press, 1971.

Pablo Serra, Juan. "What Is and What Should Pragmatic Ethics Be?" *European Journal of Pragmatism and American Philosophy* 2, no. 2 (2010): 1–15.

Richard, Shusterman. *Practicing Philosophy*. New York and London: Routledge, 1997.

Simmons, A. John. "Ideal and Nonideal Theory." *Philosophy & Public Affairs* 38 (2010): 5–36.

Singer, Peter. "Famine, Affluence, and Morality." *Philosophy and Public Affairs* 1, no. 3 (1972): 229–243.

Svoboda, Toby, Klaus Keller, Marlos Goes, and Nancy Tuana. "Sulfate Aerosol Geoengineering: The Question of Justice." *Public Affairs Quarterly* 25, no. 3 (2011): 157–179.

Svoboda, Toby. "Solar Radiation Management and Comparative Climate Justice." In *Climate Justice and Geoengineering: Ethics and Policy in the Atmospheric Anthropocene*, edited by Christopher J. Preston. Lanham: Rowman & Littlefield International, 2016, 3–14.

Svoboda, Toby. *The Ethics of Climate Engineering: Solar Radiation Management and Non-Ideal Justice*. New York: Routledge, 2017.

Svoboda, Toby, Peter J. Irvine, Daniel Callies, and Masahiro Sugiyama. "The Potential for Climate Engineering with Stratospheric Sulfate Aerosol Injections to Reduce Climate Injustice." *Journal of Global Ethics* 14, no. 3 (2018): 353–368.

7 Why Not Optimism?

This chapter offers an attack on what we may call ecological optimism. So far, I have argued that its opposite, ecological pessimism, should be accepted. It will be helpful to offer a direct account of why optimism should be rejected. Like its pessimistic counterpart, ecological optimism is a cognitive attitude but one that expects a future free of ecological catastrophe. Such a person thinks that catastrophe is unlikely to occur. There are many reasons one might think this, as we shall see. To be fair, the optimist can allow that the future will be imperfect, as there may be various ills in the future that fall short of full-blown catastrophe. He need not be like Voltaire's Pangloss. The optimism that I critique here is of a modest, more plausible form. It would be easy enough to critique someone who thinks this world is a paradise and that future perfection is a sure thing, but it will be more useful to argue against the kind of view that some people actually hold.

The Implausibility of Optimism

The most straightforward complaint against ecological pessimism is that it simply is not plausible. As we have seen, there is good evidence that we face various risks of phenomena that are plausibly deemed catastrophic: mass extinction, ocean acidification, dangerous climate change, nuclear warfare, and so on. As we have also seen, humanity is doing relatively little to remove or reduce these risks. Although it is possible that we might change our ways in the future, there is little reason to expect this, given both our track record and our failure to provide future generations with adequate resources to conduct ecological risk reduction. Let me explain both of these points.

First, as I have argued throughout this book, humanity's history is, from any reasonable moral point of view, an ugly one. On the plausible assumption that the future will resemble the past unless there is some clear reason to think otherwise, humanity's future will be morally ugly as well. As it

DOI: 10.4324/9781003570523-8

happens, making substantial progress in averting ecological catastrophe would require a serious moral commitment, for it would involve making present sacrifices (e.g., paying a carbon tax) in order to benefit persons we do not know due to their spatial or temporal distance. We are certainly capable of making such sacrifices, but we rarely do so. If one doubts this, simply look around at the social-political conditions of the present (e.g., widespread hostility toward emissions mitigation) or the past (e.g., general support or indifference regarding slavery). We are simply not a morally serious species, as I have argued elsewhere (Svoboda 2022). Hence my appeal to our track record. In order to be plausible, ecological optimism would require a fundamental moral transformation. But there is no good reason to expect such a transformation. So ecological optimism is not plausible. Moreover, if I am correct about our track record, that places the burden of proof on the ecological optimist. Continuing on our current trajectory invites catastrophe. It is possible that we will abandon this trajectory in favor of a more promising one, but it is incumbent on the optimist to provide good reasons for expecting such a change of course.

Now it might be argued that I am begging the question here, for I am merely assuming a pessimistic point in order to argue against ecological optimism. But there is an important distinction to be made between moral pessimism in general and ecological pessimism in particular. Although I believe the two to be connected, there is no conceptual link. It is possible that moral pessimism could be true while ecological pessimism remains false. For example, had conditions in the world been different, averting ecological catastrophe might have been in our immediate self-interest. In that case, our morally unserious species might have taken steps to avert catastrophe, as this would not have required any sacrifice on our own part. Of course, that possible world is not the actual one. As a matter of fact, our moral unseriousness does pose a massive obstacle to ecological risk reduction. Pointing that out is not a case of begging the question, because the morally pessimistic claim is independent of the ecologically pessimistic claim.

Second, I claimed that the present generation of humanity is failing to provide future generations with adequate resources to avert ecological catastrophe. The idea here is that any serious attempt to avert catastrophe must be an intergenerational endeavor. This is for the simple reason that many risks of potential catastrophe stretch into the future, sometimes the very distant future. Averting dangerous climate change, for example, requires long-term cooperation among many generations.[1] It is not enough for one or two generations to reduce their emissions substantially, as the next generation might simply undo all that work through its own profligacy. Of course, there is no way for the present generation to compel future generations to act, short of fanciful scenarios. Likewise, and more to the point I wish to make, there is no way for future generations to compel past generations to act, for obvious reasons. This makes it trivial for earlier generations to leave later generations high and dry. The former might simply do nothing about climate change, instead pursuing irresponsible consumption, new types of warfare, and pointless culture wars as the world burns. In that case, later generations might face dangerous climate change without the means to avert catastrophe for themselves. Of course, climate change is but one example. A very similar claim can be made for other potential catastrophes, such as mass extinction through habitat loss or ocean acidification. Inadequate preparations now may leave future generations ill-equipped to avoid catastrophe.

To be fair, some preparations are being made, but they are very modest: non-binding pledges to cut emissions, promises to provide funding for adaptation, information-gathering by the IPCC, and so on. These efforts are not enough, of course. If we wanted, we could drastically cut emissions, set aside substantial funds for future adaptation, create economic frameworks incentivizing the abandonment of fossil fuels, and invest heavily in research and development of new technologies. I am aware that such an aggressive response to climate change is not politically feasible and is unlikely to happen. That is my point. Because of this failure, future generations are unlikely to be equipped to avert, or respond to, climate catastrophe. This is another reason, distinct from our track record, for finding ecological optimism implausible. For the future to be free of catastrophe, earlier generations must make certain preparations. But we are not making those preparations. So ecological optimism is not plausible.

Let me offer a more direct argument for thinking ecological optimism implausible. I claim that, in order to avoid ecological catastrophe in the future, we would need to be extremely, implausibly lucky. For any unit of time, there is a probability of *some* catastrophic event's occurrence. Given sufficient opportunity, say because we do little to reduce the relevant risks, it becomes more and more likely that this event will occur at some point, although we cannot predict precisely when. Consider an analogy. For each year (or day, or month, or decade), there is some probability of a large-scale nuclear exchange. That probability is presumably low, but it is not zero. As noted in Chapter 1, there have been close calls in the past (Tertrais 2017). We re-run that risk each year. Over time, it becomes much more likely that this event will occur in *some* year. For example, the probability of a nuclear exchange in 2089 may be relatively low, but the probability of a nuclear exchange by (say) 2300 may be relatively high. Similarly, while each motorist runs a fairly low risk of death due to a traffic accident, it is virtually certain that some motorist will die due to an accident. The application of ecological risks is obvious. While any specific ecological catastrophe at a given point in time may be unlikely, the probability of *some* ecological catastrophe occurring at *some* point in the future is relatively much higher. This is because there are many potential catastrophes and many points at which they might occur in the indefinite future. To rely on a metaphor, if we roll the dice over and over, we are bound to hit upon a bad roll eventually, even if we have no way of predicting precisely what the bad roll will be nor when precisely it will occur.

There is a potential way out for the ecological optimist. She might claim that the probability of each catastrophe is extremely low, such that *any* ecological catastrophe is unlikely to occur in the future, despite the many opportunities. Similarly, if I roll a million dice at once, I am unlikely to see any particular roll (e.g., a million sixes), even if I spend my entire life at the table. This will not work, however. While the probability of rolling a million sixes may as well be zero, the probability of ecological catastrophe is appreciably higher. First, catastrophic phenomena are not merely theoretical. They have occurred at least several times in earth's history, such as the Great Oxidation Event and the Cretaceous–Paleogene extinction. Second, observations suggest that some catastrophes may already be occurring, as with widespread species extinction (Rounsevell et al. 2020). Third, projections (e.g., climate models) show many non-negligible risks that could well prove catastrophic, such as low-mitigation scenarios resulting in high-temperature increases (Weitzman 2009). For these reasons, the possibility of ecological catastrophe is not like the possibility of rolling a million sixes at once. The former is much more likely than the latter. This would undercut the optimist's attempt to sidestep my argument, for many of the relevant risks are non-negligible. In order to avoid ecological catastrophe, we would need matters to break our way many times and in many cases over the next several centuries or millennia. That is not impossible, but it is certainly implausible. This is why avoiding ecological catastrophe would require incredible luck, making the optimistic stance implausible.

The Dishonesty of Ecological Optimism

The foregoing section makes an epistemic case against ecological optimism. Such optimism also runs the risk of being dishonest. In saying that ecological optimism is dishonest, I am not accusing specific individuals of lying about our ecological prospects. Although that clearly happens, I addressed that issue in the earlier chapter on evil and climate obstructionism. Now I wish to examine subjects who accept, in some sense, the claim that ecological catastrophe is unlikely. Although it is possible that someone might advocate optimism as a kind of "noble lie" in service to some ulterior purposes, the dishonesty I have in mind is usually inchoate. Moreover, we are often dishonest with ourselves. Sometimes we prohibit ourselves from certain thoughts, perhaps for social or moral reasons, but without acknowledging those actual reasons. Nearly every time I mention ecological pessimism, there is immediate resistance. I am told that we must remain hopeful and never give up the fight. This is troubling not because I think hope and determination to be mistaken, but rather because the pessimistic stance is dismissed reflexively. There is no serious consideration of whether pessimism might be true, plausible, or at least reasonable. Instead, it is simply rejected for non-epistemic reasons, as if it were too dangerous even to entertain. I fear that this indicates a general bias toward optimism. As it happens, we do observe a general "optimism bias" in many humans (Sharot 2011). This was especially evident during the COVID-19 pandemic, such as by underestimating one's own susceptibility to the disease (Park et al. 2021).

Ecological optimism is often dishonest in the sense that it disregards the many indicators of catastrophe to which we have access. It does not honestly "see" this evidence, instead disregarding it. Strictly speaking, a proposition, such as "ecological catastrophe is unlikely to occur in the future," cannot be dishonest, but only true or false. Dishonesty arises as a result of a subject's relation to a proposition, as when one asserts a proposition for which she knowingly has no evidence. When I say that ecological optimism is dishonest, I mean that it is usually presumed without a serious acknowledgment of the vast evidence to the contrary. My claim here is not that ecological optimism is false, although I think it is, but rather that the optimist (usually) fails to take account of reality. It is like someone who refuses to accept the self-destructive behavior of a family member, ignoring the evidence and preferring to believe a comforting falsehood. That is understandable, but it is clearly dishonest. As with the self-destructive family member, dishonesty about our ecological prospects is not a trivial or harmless matter. Instead, it runs the risk of creating false hope and unrealistic expectations, which I discuss below.

Now it is possible that someone might be a perfectly honest ecological optimist, a person who acknowledges the evidence and nonetheless reaches the conclusion that catastrophe is unlikely to occur. Possibly he thinks the available scientific evidence has been misinterpreted, that a divine being will intercede to prevent catastrophe at the last moment, or something else. This person may be mistaken and perhaps naive, but he is not guilty of dishonesty. In many cases, we thankfully cannot know the internal psychology of other people, so it may be practically impossible to determine whether any given individual is dishonest in this way. However, I think few ecological optimists match this description. Even at the present time, many millions of people in the United States say that climate change is not a major threat (Tyson et al. 2023). Are we to believe that most of them have made a serious study of the evidence and simply formed an honest, if unorthodox, opinion? To be sure, there is a great deal of ignorance about climate change, partly because some interests have worked very hard to create and maintain that ignorance (Oreskes and Conway 2011). Again, I do not claim that such respondents are telling lies, but lying is merely one type of dishonesty. What is dishonest here is to hold an opinion despite relevant ignorance. Surveys typically include an optional response of "I don't know" or "no opinion." Few respondents opt for that, despite it being a more honest response in many cases.

Finally, ecological optimism gives rise to false hopes and unrealistic expectations (see Nguyen, 2024). Although some forms of hope may be reasonable (Shockley 2022; Thompson 2010), others are surely not. By painting an implausibly attractive picture of the future, the optimist promises too much. Those who buy into such optimism may find themselves disappointed, perhaps dangerously so. Suppose someone purchases coastal property at an amazing discount, confidently optimistic that sea-level rise will pose no problem. That may turn out to be a poor investment. Or imagine a populace that is taught to reject the supposed alarmism of pessimists, comforting itself in the belief that all will turn out well. The awakening that eventually comes due to reality's intrusion may be more painful than it would have been under a more sober assessment. Schopenhauer observes that optimism can be a cause of suffering:

> The fool runs after the pleasures of life and sees himself cheated; the sage avoids evils. for all pleasures are chimerical, and to mourn over missing out on them would be petty, indeed ridiculous. The failure to recognize this truth, encouraged by optimism, is the source of much unhappiness.
>
> (Schopenhauer 2014, 357)

He adds, "Those who, with too gloomy a gaze, regard this world as a kind of hell and, accordingly, are only concerned with procuring a fireproof room in it, are much less mistaken" (Schopenhauer 2014, 357). The implication is that, even if the pessimist mistakenly exaggerates the badness of the world, there is a kind of wisdom in that—they "are much less mistaken." Whereas the optimist foolishly holds out the expectation of satisfaction, happiness, or ecological stability, an expectation that is sure to be disappointed, exaggerated pessimism merely over-prepares. It would be better still not to exaggerate, of course. I have tried to argue that ecological pessimism is not an exaggerated form of doom-saying, but for those not convinced by my attempt, Schopenhauer's point is well taken.

Against Global Pessimism

Despite my sympathy for Schopenhauer, unlike him, I am not a global pessimist, although that would make for a simpler overall picture. If we should be pessimistic about everything, then we should be pessimistic about our ecological prospects. But I have argued only for ecological pessimism. I remain agnostic on whether we should be pessimistic in certain other areas. One reason for this is that I cannot accept Schopenhauer's arguments for global pessimism.

As we saw previously, Schopenhauer's pessimism about human well-being depends upon the dual threats of pain and boredom. I have some desire, which is either satisfied or not. If it is not satisfied, I suffer pain. If the desire is satisfied, however, I am not in the clear, because that satisfaction is fleeting. I might become bored with the object of my satisfied desire, and boredom is a kind of suffering for Schopenhauer. This gives rise to a new desire, which once again will be satisfied or not, terminating in either the pain of thwarted desire or the boredom of satisfied desire. It is a hallmark of human life—indeed, of all phenomena in the universe—that there is no state of lasting fulfillment. Everything in the universe strives toward some end. This includes human beings and animals in their desires, plants in their growth, and even non-living matter, such as an asteroid in motion. Yet nothing ever achieves its end. Most importantly for our purposes, human life is a constant struggle to satisfy an endless series of desires, which always ends in death. This is a pessimistic view, obviously. The only way the human being could be fulfilled or truly happy is by achieving something that is impossible, namely a stable state of lasting satisfaction that never grows boring.

The metaphysical basis for all this is Schopenhauer's notion of the Will as the thing-in-itself. Schopenhauer accepts the distinction between what Kant calls phenomena and noumena, roughly the "worlds" of experience and ultimate reality, respectively.[2] To offer a quick overview, for Kant objects of experience are given to us through "intuition," which conditions those objects in space and time. Famously, Kant claims that things-in-themselves (i.e., things as they really are independent of our experience) are neither spatial nor temporal. Rather, space and time are features provided by intuition. This intuited object is thought according to certain categories of what Kant calls the "understanding," and this allows us to subordinate intuited objects under concepts. For example, if I have a relevant concept, I might recognize a certain object of experience as a dog. This is only possible because my understanding supplies the appropriate concept, yet there would be nothing for the concept to apply to without intuition's provision of the relevant object of experience. Importantly, Kant is very clear that, at least for beings like us, all knowledge is limited to objects of experience and never extends to things-in-themselves. This is the so-called "restriction thesis," which holds that all knowledge is limited to appearances or the objects of experience that are conditioned spatiotemporally by our intuition (Ameriks 1985). This is because ours is a "sensible" intuition rather than an "intellectual" one, which is to say that our intuition supplies its own conditions in the forms of space and time. In intellectual intuition, as might be possessed by a possible divine being, one would instead "see" things-in-themselves as they really are, i.e., without adding conditions of their own. In short, an intellectual intuiter directly knows things-in-themselves (noumena), whereas a sensible intuiter knows only the appearances (phenomena) that somehow arise from things-in-themselves.

Schopenhauer is sympathetic to some parts of Kant's epistemological framework, but with the major exception that he thinks we can know things-in-themselves. Moreover, Schopenhauer thinks that the thing-in-itself is the Will, and everything we experience is but a manifestation of this Will. Of course, we do not seem to experience the Will directly. This is because, according to Schopenhauer, reality may be considered as representation or as Will, depending on what perspective we adopt. This corresponds, at least roughly, to Kant's phenomena–noumena distinction. We represent the world such that we experience various phenomena: spatiotemporal objects, individual entities, organisms, material things, other human beings, and so on. However, considered in itself (i.e., noumenally), the world is actually Will, which manifests itself in various ways through our representation and in line with what Schopenhauer calls the "four-fold root of the principle of sufficient reason" (Schopenhauer 2012). The Will is nothing but endless striving, what Dale Jacquette calls a "hungry will" (Jacquette 2005). This is the metaphysical basis for Schopenhauer's pessimism, not just for human beings but for all things. There is no prospect for the Will to be satisfied and cease striving, because then the thing-in-itself would cease to be the Will. Accordingly, human contentment is impossible not because it is difficult or due to some particular imperfection in human beings, but rather because human beings are, like everything else, just an expression in representation of this all-consuming Will.

That is what Schopenhauer contends, and although it is an interesting idea, few are likely to accept that the Will is the thing-in-itself, and for the same reason that few are likely to accept Plato's forms, Spinoza's God, or Leibniz's monads. All of these serve philosophical purposes in their respective systems of thought, but they are inherently implausible and metaphysically extravagant. To be sure, each philosopher has reasons, and sometimes even arguments, for adopting his extravagance, but all are subject to reasonable objections. In the case of Schopenhauer, it is just not clear why he is sure that the thing-in-itself is the Will. If we are good Kantians, we might claim that it is impossible to know this, but perhaps we should not be Kantians here. Allowing that the thing-in-itself is knowable, how can we know that it is the Will, which seems suspiciously anthropomorphic? In notes from 1868, Nietzsche claims that Schopenhauer's logical proofs for the existence of the Will fail. One problem he points out is that Schopenhauer seems to borrow predicates from the world of representation, applying them to the thing-in-itself. This is a questionable maneuver on Schopenhauer's part, perhaps even impermissible by the lights of his own philosophy. Even if one allows that the thing-in-itself is knowable in principle, Schopenhauer nonetheless maintains a strict divide between phenomena and noumena. What then justifies the application of predicates (e.g., oneness) in the world of representation to the thing-in-itself?

Going further, Nietzsche suggests that Schopenhauer merely guesses that the thing-in-itself is the Will, which "is created only with the aid of poetic intuition" (Nietzsche 1998, 260). It is as if

> the thinker standing before the riddle of the world simply has no means other than guessing, in the hope, that is, that a moment of genius will place on his lips the word that offers the key to that script, visible to all and yet unread, which we call the world.
>
> (Nietzsche 1998, 261)

It seems to me that Nietzsche is correct. The idea of the Will as thing-in-itself is an apt and powerful metaphor. If we look at the world, we do see a great deal of suffering, want, frustration, and the like. Everything appears mutable, subject to decay and death. Human life in particular contains a great quantity of suffering, much of it imposed by the wills of other human beings through greed, ignorance, or malevolence. As a poetic description, Schopenhauer's Will nicely describes this horrifying place. But is his doctrine true in a metaphysical sense? Probably not, but that does not render it useless. After heavily criticizing the philosophy of Henri Bergson, Bertrand Russell writes,

> Shakespeare says life's but a walking shadow, Shelley says it is like a dome of many-coloured glass, Bergson says it is a shell which bursts into parts that are again shells. If you like Berson's image better, it is just as legitimate.
>
> (Russell 1945, 810)

We cannot accept Schopenhauer's doctrine of the Will, but it provides a captivating poetic image. The Will, as he says,

> shows its different sides in the qualities, passions, errors, and excellences of the human race, in selfishness, hatred, fear, boldness, frivolity, stupidity, slyness, wit, genius, and so on. All of these, running and congealing together into a thousand different forms and shapes (individuals), continually produce the history of the great and the small worlds, where in itself it is immaterial whether they are set in motion by nuts or by crowns.
>
> (Schopenhauer 1969, 183)[3]

Fortunately, pessimists in general and ecological pessimists in particular need not follow Schopenhauer into metaphysical excess. For example, we might agree with Schopenhauer's pessimistic analysis of human life, even while rejecting the underlying metaphysics Schopenhauer used to explain it. We might content ourselves with the purely empirical elements of his account, shrugging at the supposed need for a deeper, metaphysical explanation. In this approach, we can agree with Schopenhauer's diagnosis of human life as subject always to either frustration or boredom, both of them, forms of discontentment. This need not be a manifestation of the Will, but merely a fact about human nature, allowing the pessimist to escape Schopenhauer's metaphysical shadow.

This purely empirical version of Schopenhauer's pessimism seems plausible enough when it comes to the pursuit of some desires, particularly those tied to our animal nature. We have desires. We either sate them or not. If the latter, we suffer. If the former, we soon become bored and turn our attention to a different desire, which is either sated or not. This is an endless cycle. Eating or drinking brings pleasure, but that soon fades and is replaced by a new desire for food or drink. There is no such thing as a meal so good that it brings perpetual satisfaction of hunger. If we could somehow satisfy each desire the moment it arises, then we would experience constant boredom, which is its own kind of suffering. If all pursuits were like this, then Schopenhauer would have a strong case for his general pessimism.

But what of pursuits that John Stuart Mill would identify as "higher pleasures," such as intellectual, interpersonal, or spiritual ventures (Mill 2002)? Schopenhauer must hold that even these fall prey to the same logic as animalistic pleasures, ending in either frustrated desire or boredom. Suppose I attempt to write a novel. The process might be painful and end in failure, but alternatively, I might enjoy the process and see the work to completion. A Schopenhauerian pessimist might say that the latter case achieves a momentary fulfillment of desire, or perhaps a series of such moments, but still terminates in boredom. I will eventually grow tired of this project, either stagnating with it as I endlessly tweak sentences or chasing fulfillment through a new project. This is why Janaway identifies Schopenhauer as a hedonist:

> "An undefended assumption in his argument is a stark form of hedonism: something adds positive value to life if and only if it involves a felt pleasure, while something contributes negative value if and only if it involves a felt pain.
>
> (Janaway 1999, 56)

But must this be the case? The strongest consideration against Schopenhauer's account comes from reflecting on the phenomenology of certain desires and attitudes. Is it not the case that we sometimes experience a stable, persistent form of satisfaction? Suppose someone takes comfort from the fact that she has written a novel, even if it remains unpublished and unread. Perhaps someone finds a kind of peace in having completed a difficult yet worthwhile task, such as sticking to a demanding exercise regimen. One might feel pride, or any of a range of other attitudes, regarding one's students or children. The Epicureans thought it possible to experience the simple joy of existing, which involves contemplating the absence of pain in the body and distress in the soul. The Stoics believed that virtue was sufficient for flourishing. Even for those pleasures that are transitory, through memory I can recall them for the rest of my life, appreciating them once again. Cases like this appear to offer counter examples to Schopenhauer's account. If we think about what it is like to experience such things, it sometimes feels as if we experience more than the fleeting pleasure Schopenhauer identifies, although I make no claim about how common this is. The implausibility of Schopenhauer's account arises from the fact that it forecloses the very possibility of such persistent satisfaction. Phenomenologically, it certainly seems that we can experience such satisfaction, even if it is rare.

A proponent of Schopenhauer's view might answer this in the following way. First, recalling Janaway's claim about Schopenhauer's hedonism, perhaps satisfaction may only come in the form of a "felt pleasure." In that case, although it is true that I may take comfort in having written a novel or feel pride about my children, the good in such things is merely the feeling of pleasure that they provide. This would render the "higher" forms of satisfaction little different from the "lower" forms. In both cases, we have a felt pleasure. Such a feeling cannot last forever. Eventually, I will become bored with it. I shall then need to find something else to satisfy me. I will either find something or not. Arguably, then, the pessimistic cycle holds even for non-animalistic forms of satisfaction. But why be pessimistic about this cycle? Perhaps it is true that I will always be at risk of either frustration or boredom, but even Schopenhauer seems to admit that temporary satisfaction is possible. If I am lucky, things might run as follows. I satisfy some desire, which brings pleasure for a while. Before it fades into boredom, I find another desire to satisfy, which again brings pleasure for a while. I repeat this process indefinitely. Why shouldn't this indefinite series of satisfied desires count as a happy life? It may be that no single satisfaction is persistent throughout my life, but as long as I have many satisfactions in uninterrupted succession, so what?

Of course, matters are more complicated. At any time, we will have many desires that need to be satisfied, and it is very unlikely that we will be able to satisfy them all across our lives. While it may be conceivable that someone should enjoy uninterrupted pleasure through the successive satisfaction of all desires, that will not happen for real human beings. Even for the luckiest among us, there will be frustration and boredom—in short, pain. But why not accept this reality and insist that human beings can have a good life if on balance their pleasure substantially outweighs their pain? I might claim to be happy, to be leading a good life, because most of my desires, especially the most important ones, are satisfied the vast majority of the time. Schopenhauer needs to claim, implausibly to my mind, that these many and consistent satisfactions do not compensate for the occasional disappointment (Simmons 2021).

Fortunately, my position does not require global pessimism. The ecologically pessimistic attitude can stand on its own. In my view, our ecological future is likely to be bleak. There is no necessity for this. It is just an outcome of humanity's recent capacity to impact nature and its relative indifference to suffering and injustice. So we may be ecological pessimists without accepting the less plausible aspects of pessimism in general.

Notes

1 In regard to climate change specifically, this is related to what Gardiner calls the "pure intergenerational problem" (Gardiner 2006, 2011). Briefly put, climate progress requires multiple generations to cooperate in mitigating emissions. Each generation is faced with the choice of whether to cooperate or not. It appears "rational" for each generation to defect, however. For one thing, a current generation might be a free-rider, enjoying the benefits of past generations' mitigation without contributing anything itself, letting others do the work. For another, a current generation might observe that there is no guarantee that future generations will choose to cooperate and therefore decide not to run the risk of making a pointless sacrifice by cutting its own emissions. Of course, if each generation decides not to cooperate, then no generations will cooperate, in which case dangerous climate change will occur.
2 We might think of the phenomena–noumena distinction as specifying two "aspects" of reality rather than two "worlds," but that is not particularly important to the points I am making here. See Allison (2004).
3 I quote here from Payne's translation on account of its readability.

References

Allison, Henry. *Kant's Transcendental Idealism: An Interpretation and Defense.* New Haven: Yale University Press, 2004.
Ameriks, Karl. "Hegel's Critique of Kant's Theoretical Philosophy." *Philosophy and Phenomenological Research* 46, no. 1 (1985): 1–35.
Gardiner, Stephen M. "A Perfect Moral Storm: Climate Change, Intergenerational Ethics and the Problem of Moral Corruption." *Environmental Values* 15, no. 3 (2006): 397–413.
Gardiner, Stephen M. *A Perfect Moral Storm.* New York: Oxford University Press, 2011.
Jacquette, Dale. *The Philosophy of Schopenhauer.* Montreal: McGill-Queen's University Press, 2005.

Janaway, Christopher. "Schopenhauer's Pessimism." *Royal Institute of Philosophy Supplements* 44 (1999): 47–63.

Mill, John Stuart. *Utilitarianism: And the 1868 Speech on Capital Punishment.* Indianapolis: Hackett, 2002.

Nguyen, Anh-Quân. "Pessimism for Climate Activists." *Ethics & the Environment* 29, no. 1 (2024): 109–137.

Nietzsche, Friedrich. "On Schopenhauer." In *Willing and Nothingness*, edited by Christopher Janaway. New York: Oxford University Press, 1998.

Oreskes, Naomi, and Erik M. Conway. *Merchants of Doubt: How a Handful of Scientists Obscured the Truth on Issues from Tobacco Smoke to Global Warming.* USA: Bloomsbury Publishing, 2011.

Park, Taehwan, Ilwoo Ju, Jennifer E. Ohs, and Amber Hinsley. "Optimistic Bias and Preventive Behavioral Engagement in the Context of COVID-19." *Research in Social and Administrative Pharmacy* 17, no. 1 (2021): 1859–1866.

Tyson, Alec, Cary Funk, and Brian Kennedy. *What the Data Says about Americans' Views of Climate Change.* Pew Research Center, 2023. https://www.pewresearch.org/short-reads/2023/08/09/what-the-data-says-about-americans-views-of-climate-change/.

Rounsevell, Mark D. A., Mike Harfoot, Paula A. Harrison, Tim Newbold, Richard D. Gregory, and Georgina M. Mace. "A Biodiversity Target Based on Species Extinctions." *Science* 368, no. 6496 (2020): 1193–1195.

Russell, Bertrand. *A History of Western Philosophy.* New York: Simon & Schuster, 1945.

Schopenhauer. *The World as Will and Representation, Volume 1.* Translated by E. F. J. Payne. New York: Dover, 1969.

Schopenhauer, Arthur. *On the Fourfold Root of the Principle of Sufficient Reason and Other Writings.* Translated by Christopher Janaway, David E. Cartwright, and Edward E. Erdmann. Cambridge: Cambridge University Press, 2012.

Schopenhauer, Arthur. "Aphorisms on the Wisdom of Life." In *Parerga and Paralipomena, Volume 1.* Translated and edited by Sabine Roehr and Christopher Janaway. Cambridge: Cambridge University Press, 2014.

Sharot, Tali. "The Optimism Bias." *Current Biology* 21, no. 23 (2011): R941–R945.

Shockley, Kenneth. "Living Well Wherever You Are: Radical Hope and the Good Life in the Anthropocene." *Journal of Social Philosophy* 53, no. 1 (2022): 59–75.

Simmons, Byron. "A Thousand Pleasures Are Not Worth a Single Pain: The Compensation Argument for Schopenhauer's Pessimism." *European Journal of Philosophy* 29, no. 1 (2021): 120–136.

Svoboda, Toby. *A Philosophical Defense of Misanthropy.* New York: Routledge, 2022.

Tertrais, Bruno. "'On the Brink'—Really? Revisiting Nuclear Close Calls Since 1945." *The Washington Quarterly* 40, no. 2 (2017): 51–66.

Thompson, Allen. "Radical Hope for Living Well in a Warmer World." *Journal of Agricultural and Environmental Ethics* 23 (2010): 43–59.

Weitzman, Martin L. "On Modeling and Interpreting the Economics of Catastrophic Climate Change." *The Review of Economics and Statistics* 91, no. 1 (2009): 1–19.

Index

For Product Safety Concerns and Information please contact our EU
representative GPSR@taylorandfrancis.com
Taylor & Francis Verlag GmbH, Kaufingerstraße 24, 80331 München, Germany

www.ingramcontent.com/pod-product-compliance
Ingram Content Group UK Ltd.
Pitfield, Milton Keynes, MK11 3LW, UK
UKHW022325100726
473146UK00010B/636